SUNDIALS

THEIR THEORY
AND CONSTRUCTION

Medieval noon mark from Bedos de Celles, *La Gnomonique pratique* (Paris, 1790).

SUNDIALS

THEIR THEORY
AND CONSTRUCTION

ALBERT E. WAUGH

Dover Publications, Inc.

NEW YORK

Published in Canada by General Publishing Company, Ltd., 30 Lesmill Road, Don Mills, Toronto, Ontario.
Published in the United Kingdom by Constable and Company, Ltd., 10 Orange Street, London WC 2.

Sundials: Their Theory and Construction is a new work, first published by Dover Publications, Inc. in 1973.

International Standard Book Number: 0-486-22947-5
Library of Congress Catalog Card Number: 73-76961

Manufactured in the United States of America
Dover Publications, Inc.
180 Varick Street
New York, N.Y. 10014

O God! methinks it were a happy life,
To be no better than a homely swain;
To sit upon a hill, as I do now,
To carve out dials quaintly, point by point,
Thereby to see the minutes how they run,
How many make the hour full complete;
How many hours bring about the day;
How many days will finish up the year;
How many years a mortal man may live.

Shakespeare
King Henry VI, Part III, 2, v.

Preface

The origin of sundials is lost in antiquity. The student of the history of science finds in them evidence of very early understanding of fundamental relationships of astronomy. The modern science teacher uses them to illustrate vividly the first principles of the solar system. The artist finds examples of medieval design and craftsmanship in their intricate engraving or stonework. The architect uses them to good advantage to grace a public building, and the landscape architect finds few things more suitable as accents for a formal garden. The student or teacher of high school mathematics finds here a fertile field for applying the elementary theorems of plane and spherical trigonometry. The teacher of manual arts finds in sundials the basis for projects ranging from those which he can confidently assign to a beginner to those worthy of the talents of the most gifted and skilled. Boy Scouts and 4-H Club boys find the sundial an interesting and instructive project well within their capabilities. What better project is there for a summer camp than laying out a large analemmatic dial on the parade ground? Work with sundials can be simple enough to match the budding talents of a twelve-year-old; yet we know that sundials attracted the mature interests of such prodigies as Sir Isaac Newton, Sir Christopher Wren and Thomas Jefferson. No one need feel that the subject is beyond him; no one should consider it unworthy of his talents.

This book is planned to meet the needs of readers of all ages and backgrounds. Simple mathematical methods are given for those who want them, but the reader who fears the numerical approach may skip them, using instead the graphical approaches which do not require even addition or subtraction. And for the most part, the reader may choose his own starting point, beginning at once with the chapter which describes the type of dial which interests him.

My own library contains well over one hundred old books on sundials, many of them beautifully bound and illustrated. I am deeply indebted to these authors of yesteryear and have made frequent reference to them. I can well take as my own the disclaimer which Geoffrey Chaucer wrote in the Prologue of his *Treatise on the Astrolabe* nearly 600 years ago (as transposed roughly into modern English):

> Keep in mind that I lay no claim to having discovered these things through my own skill. I am but an ignorant compiler of the works of ancient astronomers, and have but put their material into my own words for your instruction; and with this sword shall I slay envy.[1]

Edwin Burt and Malcolm Gardner have given me invaluable assistance in collecting my sundial library, and I was fortunate in being able to add a number of items from Mr. Gardner's private collection to my own after his death. Edwin Pugsley has been a source of constant inspiration to me, and was generous enough to make the gnomons for two of my large public dials. His summer "astrolabe parties" have brought together a congenial group of aficionados. Kenneth Lynch, the master craftsman, has thrilled me with his magnificent creations. My wife's patient good nature, which I have tested so often and so sorely, is evidenced by the fact that she suffered me to make an analemma by pounding long lines of copper-headed tacks into the hardwood floor of her front hall! "What a strange obsession," she remarks, "for a grown man who owns a watch!"

Grandpa's Farm   ALBERT E. WAUGH
Storrs, Connecticut
September 28, 1972

---

[1] "But considre wel that I ne usurpe not to have founden this werk of my labour or of myn engyn. I n'am but a lewd compilator of the labour of olde astrologiens, and have it translatid in myn Englissh oonly for thy doctrine. And with this swerd shal I sleen envie."

Table of Contents

Latitudes and Seasons. Table A.9. Data for Laying Out Dials in a Unit Square. Table A.10. Hour Angles for Horizontal and Vertical Direct South Dials. Table A.11. Factor Table for Reflected Ceiling Dials. Table A.12. Factor Table for Finding Distances of Hour Lines from the Equinoctial on Reflected Ceiling Dials.

1
Historical Sketch

Fortunately we need not start by defining time, since the concept has perplexed philosophers and lexicographers and served as the basis for learned and inconclusive arguments. We shall assume, with the man in the street, that we know what time is, and that our problem is measuring it rather than defining it.

Very early in human history men must have recognized the passage of time. Its major subdivisions were marked by the sequence of day and night and by the passage of the seasons. Timekeeping at this level involved merely counting the days or the years—or, with the American Indian, the cycles of moon phases. There were no such obvious subdivisions of the day, and no one knows when men first began to count the hours, nor what they used to measure their passage. It is certain that one of the early methods involved observations of shadows. As the hours of any day passed it was apparent that the shadows changed slowly in their direction and in their length. The shadows of early morning were long, and stretched toward the west. As noon approached the shadows grew shorter and swung into the north.[1] Then through the afternoon the shadows lengthened again and reached toward the east. The hour of the day could be estimated from either the length or the direction of the shadow.

At first the day was apparently merely divided into three parts— morning, afternoon and night—by the three phenomena of dawn, noon and sunset. Dawn and sunset were obvious, and noon was the moment when the shadows were shortest for the day. Later, men noted the change in length of shadows more carefully, and judged the time of day roughly by measuring their own shadows—stepping them off with their own feet, "heel to toe." The Venerable Bede

[1] In the Southern Hemisphere the noon shadows lie toward the south.

gave a table[2] about 700 A.D. for use in telling the time of day by this method (Table 1.1).

TABLE 1.1
LENGTH OF ONE'S SHADOW IN "FEET" AT VARIOUS HOURS OF THE DAY AT VARIOUS TIMES OF THE YEAR

hour of the day	Jan. Dec.	Feb. Nov.	Mar. Oct.	Apr. Sept.	May Aug.	June July
1 or 11	29	27	25	23	21	19
2 or 10	19	17	15	13	11	9
3 or 9	17	15	13	11	9	7
4 or 8	15	13	11	9	7	5
5 or 7	13	11	9	7	5	3
6	11	9	7	5	3	1

In interpreting this table one must understand that in Bede's day men counted the hours from dawn, so that the hour of "3" means "the end of the third hour after dawn." The time from dawn to sunset was divided into 12 equal "hours," but since the time from dawn to sunset was longer in summer than in winter, the "hours" of summer were also longer than the "hours" of winter. For many centuries these "unequal hours" or "temporary hours" were used over much of the earth. The "hours" of any one day were equal, but the "hours" of winter were short and the "hours" of summer long. It is for this latter reason that we refer to them as "unequal hours."

We shall have occasion to speak further of these old unequal hours, but the modern reader may find it amusing to experiment with a roughly comparable table computed for modern hours and for the latitude of New York City or Chicago (Table 1.2). If you live fairly close to this latitude you may test the table by stepping off the length of your own shadow with your own feet and comparing the time estimated from the table with that shown on your watch.

Chaucer, who wrote his *Canterbury Tales* about 1390 or 1400 A.D., gives at least two illustrations of this method of telling the time of day. In the opening lines of his "Parson's Prologue" he says:

> It was four o'clock according to my guess,
> Since eleven feet, a little more or less,
> My shadow at the time did fall,
> Considering that I myself am six feet tall.

[2] Reproduced in F. K. Ginzel, *Handbuch der mathematischen und technischen Chronologie* (Leipzig: 1914) Vol. III, p. 88.

TABLE 1.2
LENGTH OF ONE'S SHADOW IN "FEET" ON THE 22ND DAY OF VARIOUS MONTHS
IN LATITUDE 41° [3]

hour of the day	Dec.	Jan. Nov.	Feb. Oct.	Mar. Sept.	Apr. Aug.	May July	June
noon	13	11	8	5	3	2	2
11 or 1	14	12	9	6	4	3	3
10 or 2	17	14	10	7	5	4	4
9 or 3	26	21	14	10	7	6	5
8 or 4	70	47	24	15	10	8	8
7 or 5				30	18	13	12
6 or 6					46	26	22
5 or 7							72

And near the opening of the Introduction to his "Man of Law's Tale" he tells us:

> ... the shadow of each tree
> Had reached a length of that same quantity
> As was the body which had cast the shade;
> And on this basis he conclusion made:
> ... for that day, and in that latitude,
> The time was ten o'clock ....

But in many ways the direction of a shadow is a more satisfactory time-teller than its length. Boy Scouts are told that they can tell the direction from their watches. They are instructed to hold the watch face upwards and point the hour hand toward the sun. The south point will then lie, it is said, half way between the hour hand and 12 o'clock. This rule is actually very rough, but perhaps better than none at all.

We do not know when men first began to use instruments which were at all similar to modern sundials. A stone fragment in a Berlin museum is thought to be the earliest known sundial, dating from about 1500 B.C. The Bible mentions what some authorities take to have been a sundial (although the meaning is by no means certain) in the days of Ahaz, king of Judah some 700 years before Christ.[4] About a century later the Greek philosopher and astronomer Anaximander of Miletus is said to have introduced the sundial into Greece. Herodotus, who lived in Asia Minor and Greece about

---

[3] Times given in this table are local apparent times. Computations are based on the assumption that a man's height is just six times the length of his own foot.

[4] The so-called dial of Ahaz. See Isaiah 38:8 and II Kings 20:11.

450 B.C., tells us that "It was from the Babylonians that the Greeks learned about the pole, the gnomon and the twelve parts of the day"; and sundials had become so common in Rome by 200 B.C. that the comic dramatist Plautus condemned in verse "the wretch who first . . . set a sundial in the market place to chop my day to pieces." Vitruvius, a contemporary of Julius Caesar, bemoaned the fact that he could not invent new types of sundials, since the field was already exhausted. He lists a dozen or more types, giving the names of their inventors. We do not know anything about the appearance of many of these early dials, and cannot guess the degree of their accuracy.

Many medieval English churches carry what appear to be crude sundials cut or scratched directly into the stone of their walls. These appear to have been used primarily to note the times of the prayers. One of these dials, at Kirkdale in Yorkshire, carries an inscription in Old English which reads in part, "This is the day's sun-marker at every tide."[5] This will be understood only if we realize that the Saxons divided the day not into hours, but into "tides"—from which we still get such words as "noontide" and "eventide."

Sometime and somewhere—no one knows when or where—it was discovered that the shadow cast by a slanting object might be a more accurate timekeeper than the shadow cast by a vertical one. If, in fact, the shadow-casting object was parallel to the earth's axis, the direction of its shadow at any given hour of the day was constant regardless of the season of the year. It has been suggested[6] that this discovery occurred in the first century A.D., but be that as it may, men had now discovered the system which remained the principal basis for time-telling for nearly thirteen centuries. In fact, sundials remained in use long after the invention of the clock, since early clocks were erratic and needed frequent correction by the sundial. Our frontispiece reproduces an old print showing three gentlemen waiting to set their watches when the sun dial shows that the moment of noon has arrived,[7] and many a New England housewife paced her morning's chores with the movement of the shadows across the kitchen floor. While men have used many other means of telling time—sandglasses, waterclocks, candles and graduated oil

[5] Arthur R. Green, *Sundials* (New York: Macmillan, 1926), pp. 14–15.

[6] R. Newton Mayall and Margaret W. Mayall, *Sundials* (Boston: Hale Cushman & Flint, 1938), p. 15.

[7] Dom Francois Bedos de Celles, *La Gnomonique pratique* (Paris: 1790).

lamps (or else relied on the crowing of cocks and other natural phenomena), nevertheless for at least ten and perhaps twenty centuries the sundial was the major timekeeping device used by man.[8]

---

[8] Derek Price cautions us (in *Technology and Culture*, Vol. V, No. 1, 1964) that "It would be a mistake to suppose that ... sundials ... had the primary utilitarian purpose of telling the time. Doubtless they were on occasion made to serve this practical end, but on the whole their design and intention seems to have been the aesthetic or religious satisfaction derived from making a device to simulate the heavens. Greek and Roman sundials, for example, seldom have their hour-lines numbered, but almost invariably the equator and tropical lines are modelled on their surfaces and suitably inscribed. The design is a mathematical *tour-de-force* in elegantly mapping the heavenly vault."

2
Kinds of Time

American newspapers on July 21, 1969, featured accounts of man's first landing on the moon, and at various places they gave the time of Neil Armstrong's first step on the moon as:

10:56 P.M. Eastern Daylight Time on the 20th.

9:56 P.M. Eastern Standard Time on the 20th.

2:56 A.M. Greenwich Mean Time on the 21st.

3:56 A.M. British Summer Time on the 21st.

Here were four different ways of describing the same moment of time; yet they were but four of a great many possible ways which must certainly have been used by newspapers in various parts of the world as they interpreted the news for their readers. If we are to design a sundial to "tell time," we must first decide what kind of time it is to tell.

The Sun's Apparent Motion. Every schoolchild knows that the earth revolves around the sun even though it looks as though the sun were revolving around us. For our purposes it really makes no difference, since a sundial designed to tell time on an earth with the sun revolving around it would be identical in every detail with one designed for use on an earth which was revolving around the sun. In our treatment of the matter we shall ordinarily describe things as they seem rather than as they really are. We shall thus speak of the sun's "rising in the east" and "moving across the sky from east to west" until it "sets in the west" in the evening.

Differences in Longitude. A train running from New York to San Francisco appears in Albany before it reaches Chicago, and in Chicago before it reaches Denver. Similarly as the sun moves across the sky from east to west it appears first to people living on the East Coast, and later to people living farther west. When we see the

6

sun directly south of us at midday[1] it has already passed its high point for people to our east and they see it already falling in the west, while people to our west see it still rising higher in their eastern sky. They all see the same sun at the same moment, but they see it in different directions reflecting differences in their points of view.

The *meridians* are imaginary lines running along the earth's surface from the North to the South Pole, lying everywhere exactly in a north-south direction. Our own meridian is, then, nothing more than a north-south line running through the particular spot where we happen to be. At any moment of time the sun is over one of these meridians, and everyone who is located on that meridian says that it is noon. Everyone east of that meridian says it is afternoon, and everyone to the west calls it morning. If we are to keep time by the sun we must realize that all places on the same meridian (all places due north and south of each other) will have the same time, but in all other places on the earth's surface the time will be different—later in places to the east and earlier in places to the west. Places on the same meridian are said to have the same *longitude*, and we commonly measure longitudes by their angular distances east or west from the *standard meridian* which passes through Greenwich, England. One of the earth's meridians starts (like all meridians) at the North Pole, runs due south through Greenwich, continues south across the earth's Equator, and finally reaches the South Pole. This is the standard meridian, with longitude 0°. Another meridian runs from the North Pole through New York City across the Equator to the South Pole. The arc of the Equator between these two meridians measures 73°50′, and since New York is west of Greenwich we say that the longitude of New York City is 73°50′ west. Similarly the longitude of Tokyo, Japan, is 139°45′ east of Greenwich. Since the sun makes one complete circuit of the heavens in 24 hours, passing over 360° of longitude, it obviously passes at the rate of 15° of longitude each hour, or through 1 degree each 4 minutes. If we note that the sun has covered an angle of 30° since it was last on our meridian we know that it is 2 hours past noon. The sun's angular distance from our meridian at any moment is called the *sun's hour*

---

[1] Here and hereafter we shall speak of the sun's appearance to people in north temperate climates. South of the Tropic of Capricorn the noonday sun appears in the north, and between the tropics it appears sometimes in the north and sometimes in the south and sometimes directly overhead.

angle, and as we see from the preceding examples, if the sun's hour angle is 45° west, the time is 3 P.M., while when the hour angle is 30° east it is 10 A.M.

Local Apparent Time. If we are to tell the time by the sun's hour angle, no two points will share the same time unless they lie on the same meridian with one directly north of the other. The time is thus *localized* to a particular meridian, and since it is also based on the *apparent* motion of the sun we call it *local apparent time* and often symbolize it with its initials as L.A.T. This is the kind of time shown on most sundials, and until about a century ago it was the kind of time almost universally used in daily life. Yet it suffers from disadvantages which have led most people to discard it in favor of some other kind of time.

For one thing, it is inconvenient to use a system of timekeeping which is so narrowly localized. No two places have the same L.A.T. unless they lie on the same meridian. Two cities lying 100 miles apart in an east-west direction will differ by about $7\frac{1}{2}$ minutes in L.A.T., while two towns only $13\frac{1}{2}$ miles apart will differ in L.A.T. by 1 minute. There is even a difference of about $\frac{1}{4}$ second in L.A.T. at opposite ends of a football field if it lies in an east-west direction, and precisely accurate clocks would show slightly different times in different rooms of the same house.[2]

TABLE 2.1
EAST–WEST DISTANCES OF ONE SECOND IN TIME AT VARIOUS LATITUDES

latitude:	15°	20°	25°	30°	35°	40°	45°	50°	55°
distance (ft.):	1468	1428	1378	1316	1245	1164	1075	977	872

Local Mean Time. The second disadvantage of L.A.T. arises from the fact that when we measure days by the sun they turn out to

[2] These figures are approximately correct for latitudes of the middle United States. Readers wishing to make applications in their own localities may proceed thus: At the Equator each 1520 feet in an east-west direction make a difference of 1 second in time. In other latitudes one multiplies 1520 feet by the cosine of the latitude to find the corresponding distance. Thus at a latitude of 41°, which is approximately that of New York or Chicago, the cosine of 41° being 0.7547, we multiply 1520 feet by 0.7547 and find that an east-west distance of 1148 feet makes a difference of 1 second of time. A football field of 300 feet would give a time difference of just over 0.26 seconds. At representative latitudes the east-west distances corresponding to time differences of 1 second are indicated in Table 2.1.

differ among themselves in length.[3] About Christmastime the days are about $\frac{1}{2}$ minute longer than average and in mid-September about 20 seconds shorter. These small discrepancies accumulate until in mid-February the sun reaches the meridian almost $14\frac{1}{2}$ minutes later than it would if all days were equal in length, and early in November the sun reaches the meridian about $16\frac{1}{2}$ minutes too early. These variations of 14 minutes one way and 16 minutes the other amount to just over $\frac{1}{2}$ hour, which would be decidedly inconvenient for scientific purposes, and today we would consider it unacceptable even for everyday affairs.

Instead, then, of reckoning time from the irregularly moving real sun, we usually reckon it from an imaginary *mean sun*—a fictitious heavenly body which moves in the celestial equator at a constant speed which is just equal to the average speed with which the real sun moves in the ecliptic. If the real sun and the mean sun start off together, the real sun, moving irregularly, will sometimes run ahead of the mean sun and sometimes lag behind it, but at the year's end they will finish the course together.

Time measured by the hour angle of the real or apparent sun is called *apparent time*, whereas if we measure the hour angle of the mean sun we find the *mean time*. Mean time has the advantage that it is uniform, but since the mean sun, like the real sun, is over but one meridian at a time, mean time, like apparent time, will be local. All places on a given meridian will have the same mean time, but no other places will share that time. Hence we call it *local mean time*, symbolized by L.M.T., as contrasted with L.A.T.

The Equation of Time. We are especially interested in the differences between L.A.T. and L.M.T., since most sundials, influenced by the real sun, show apparent time, while our clocks and watches, running regularly, show mean time. Thus as the real sun and the mean sun run their separate courses, sometimes one ahead and sometimes the other, our sundials and our watches will reflect the same differences. Sometimes our sundials will appear to be "fast" and sometimes to be "slow" when compared with an accurate watch. Neither the sundial nor the watch is "wrong." They merely record different kinds of time.

---

[3] The variations arise in part from the fact that the earth moves more swiftly in its orbit when close to the sun that when farther away, and in part from the fact that the sun's apparent path lies on the ecliptic rather than on the Equator.

Four times each year the real sun and the fictitious mean sun are together, and on these dates the sundial and the clock agree. While the dates of agreement vary slightly from year to year as we adjust for leap years, they fall at about April 16, June 14, September 2, and December 25. If we compare our sundial with an accurate clock we will find that from the first of the year until April 16 the dial is running behind the clock, from April 16 until June 14 the dial is ahead, from June 14 until September 2 the dial is slow again and from September 2 until Christmas the dial is fast. For the final week of the year the dial is slow again.

The amount by which the clock and the sundial differ at any given moment is called the *equation of time*. The general trend of these amounts throughout the year is shown in Figure 2.1, where we see

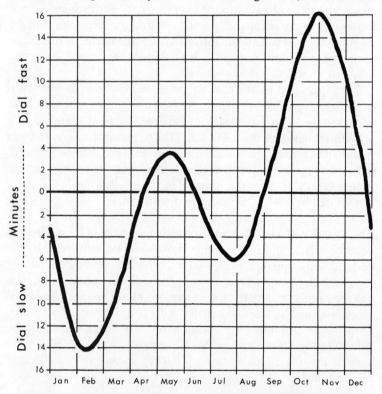

FIGURE 2.1 The equation of time, showing by how much a sundial is "fast" or "slow" at various times of year when compared with an accurate clock. For more accurate values, see Appendix, Table A.1.

the dial running slow early in the year but very fast in late October and early November, with wavy fluctuations in between. While this figure gives a good general picture, it cannot be read with much accuracy. If we want to know the value of the equation of time more precisely we can search out the line for the appropriate month on Figure 2.2 and read our values there. Even more accurate are the figures tabulated in Table A.1 of the Appendix. This appendix table

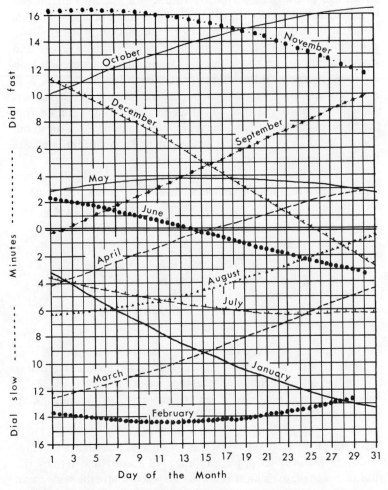

FIGURE 2.2 The equation of time. Read the value from the line for the appropriate month.

tells us, for example, that on July 4 the sundial is 4 minutes 8 seconds slow when compared with an accurate clock or watch. The values tabulated in the Appendix are averages of values which really vary only slightly from year to year, depending mainly on the proximity of the quadrennial leap-year adjustment; yet these tabulated average values are sufficiently accurate for practically all dialling pruposes and never differ from actual values by as much as a minute. The advantage of using more precise values will seldom be sufficient to offset the trouble of their calculation. Yet more precise values are available if needed. Many almanacs give the necessary information, and most national governments issue tables for surveyors and mariners from which the values may be derived. Simplest and least expensive, perhaps, is the little paper-covered *Ephemeris of the Sun, Polaris, and Other Selected Stars*, published annually by the Nautical Almanac Office of the U.S. Naval Observatory and available from the Superintendent of Documents in Washington, D.C., for about 35¢. Every dialling enthusiast will want to keep current copies of this publication on hand.

Converting from Apparent to Mean Time. If we know the value of the equation of time we can convert apparent to mean time or vice versa. Our rules are:

(1) When the sundial is "fast":
    (a) Add the equation of time to L.A.T. to get L.M.T.
    (b) Subtract the equation from L.M.T. to get L.A.T.
(2) When the sundial is "slow":
    (a) Add the equation of time to L.A.T. to get L.M.T.
    (b) Subtract the equation from L.M.T. to get L.A.T.

For example, on March 2 the sundial shows the local apparent time of 10:17 A.M. Table A.1 of the Appendix indicates that the sundial is 12 minutes and 23 seconds slow, so the local mean time is 10:29:23. Similarly at 3:30 P.M., L.M.T. on November 23 the sundial, being 13 minutes 45 seconds fast, will show 3:43:45 P.M.

Standard Time. We have noted that mean time eliminates the day-to-day inequalities of apparent time, but mean time retains the disadvantage that it is localized. No two places share the same mean time unless they are in the same longitude. In the latitude of New York City, every $13\frac{1}{3}$ miles we travel east or west brings a change of

1 minute in L.M.T. The man in Rockford, Illinois, attempting to catch a plane 70 miles away at O'Hare International Airport will miss his connection by 5 minutes if both the man and the plane are operating on their own individual L.M.T. These local variations in time brought little inconvenience when travel and communication were uncommon and slow; but the telegraph, telephone, radio, automobile and airplane changed all that. Until about a century ago each locality kept its own individual local mean time, but in 1883 the railroads of the United States adopted a system which divided the country into zones, with one single time used throughout any zone, but with adjacent zones differing one from the next by exactly an hour. In theory each zone was 15° of longitude in width, centered at longitudes 15°, 30°, 45°, 60°, 75°, etc. But a rigid adherence to this plan would bring its own inconveniences. For example, the theoretical boundary between the Eastern Standard Time zone and the zone of Central Standard Time is $82\frac{1}{2}°$ west of Greenwich. This meridian happens to run directly through the little hamlet of Berlin Heights, Ohio (and, of course, through many other places). Strict adherence to the rule would force clocks in the eastern part of town to run one hour ahead of those a few blocks farther west. A man with a dentist's appointment at 2:30 P.M. would have to find out which side of the boundary his dentist worked on. The Berlin Heights farmer, going out to milk the cows in the morning, might find an hour's difference in time between the kitchen and the adjoining milk room. Instead, then, of following the meridians exactly, the borders of the standard time belts dodge this way and that through the thickly settled parts of the world to avoid cities or to follow state and county lines in an effort to avoid dividing political units between time belts. This is evident from Figure 2.3, which shows the boundaries of the four major standard time belts of the continental United States.

Within any standard time belt all localities keep the same time—the local mean time of the belt's standard meridian. If we include the entire continent of North America we find nine standard time belts in use (Table 2.2). At sea, where there are no reasons for irregularities in zonal boundaries, the rules are applied exactly, and the resulting times are called *zone times*. *Standard times*, then, are uniform times kept throughout some designated region, being the local mean time of the standard meridian of that region, but with the boundaries of the regions adjusted to meet local conditions.

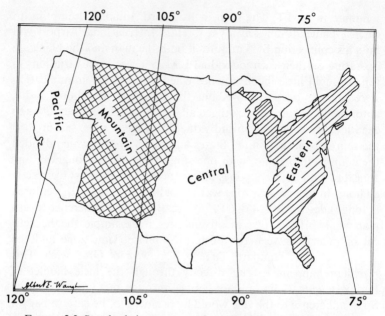

FIGURE 2.3 Standard time zones and standard time meridians in the continental United States.

Longitude Corrections. The tables showing the equation of time, such as Table A.1 of the Appendix, show the differences between the time shown on a sundial and the time shown on an accurate clock which is keeping local mean time. But most clocks keep not local

TABLE 2.2
STANDARD TIME BELTS OF NORTH AMERICA

name of standard time	standard meridian	hours earlier than Greenwich
Newfoundland	$52\frac{1}{2}°$ W	$3\frac{1}{2}$
Atlantic	60°	4
Eastern	75°	5
Central	90°	6
Mountain	105°	7
Pacific	120°	8
Yukon	135°	9
Alaska-Hawaii	150°	10
Bering	165	11

mean time, but standard time. Thus Boston, Massachusetts, is in longitude 71°05′ west, but its clocks show the local mean time not of that meridian, but of the meridian 75° west of Greenwich, since 75° is the standard meridian of the Eastern Standard Time belt. If we wish, then, to correct our sundial for comparison with an accurate watch, we must correct not only for the equation of time, but also for the difference in longitude—the difference between the longitude of the place where the dial is used and the longitude of the standard meridian of the appropriate standard time zone. Since one full revolution of the earth is 360° of longitude and corresponds to 24 hours of time, it is apparent that each hour of time corresponds to 15° of longitude, or each degree of longitude corresponds to 4 minutes of time. Each minute of longitude corresponds to 4 seconds of time, likewise. Thus if two places differ in longitude by 28°, they differ in time by 28 times 4 minutes or by 112 minutes of time or by an hour and 52 minutes. This system of conversion from longitude to time is used to find local apparent time.

Conversions from Standard Time to Local Apparent Time. These conversions involve two steps: adjustment for the equation of time, and adjustment for longitude. The problem can best be understood by an example. Suppose that on November 1 at Pittsburgh, Pennsylvania, a clock shows 3:37 P.M. Eastern Standard Time. What time should be shown by an accurate sundial which registers local apparent time? Our conversion requires two steps. First we note that the longitude of Pittsburg is 80°00′ west, and that the standard meridian of the Eastern Time zone is 75°00′ west. This is a longitude difference of 5°, and since each degree of longitude corresponds to 4 minutes of time, the 5° of longitude represent a difference of 20 minutes in local time. Our watch shows Eastern Standard Time, which is the local mean time of the 75th meridian, but Pittsburgh is 5° of longitude or 20 minutes of time west of the 75th meridian, so its local time is 20 minutes earlier than 3:37 P.M., or 3:17 P.M. But on November 1 the sundial is fast by 16 minutes and 20 seconds (we usually write this as 16^m20^s), so the sundial will be showing a time of 3:17 plus 16^m20^s or 3:33:20 P.M. We can summarize the rules for converting from standard to local apparent time or vice versa by the following tabular shorthand directions, in which L.A.T. represents the local apparent time, S.T. represents the standard time, L represents the longitude correction at the rate of 4 minutes per degree of longitude, and *e* represents the equation of time. Our rules are:

(A) If the dial is east of its standard time meridian:
  (1) If the sundial is "fast" (see Appendix, Table A.1):
    (a) L.A.T. = S.T. + L + e
    (b) S.T. = L.A.T. − L − e
  (2) If the sundial is "slow" (Appendix, Table A.1):
    (a) L.A.T. = S.T. + L − e
    (b) S.T. = L.A.T. − L + e
(B) If the dial is west of its standard time meridian:
  (1) If the sundial is "fast" (Appendix, Table A.1):
    (a) L.A.T. = S.T. − L + e
    (b) S.T. = L.A.T. + L − e
  (2) If the sundial is "slow" (Appendix, Table A.1):
    (a) L.A.T. = S.T. − L − e
    (b) S.T. = L.A.T. + L + e

These equations appear more frightening than they really are. They merely tell us whether to add or to subtract the correction factors in converting from L.A.T. to S.T. or vice versa. Taking one more example: Suppose a sundial at South Bend, Indiana, shows a time of 10:25 A.M. on April 8. What is the correct Central Standard Time? Since Central Standard Time is based on 90° longitude and South Bend is in 86°15′, the longitude difference is 3°45′ with South Bend east of the standard meridian; and on April 8 the sundial is "slow." Our rule, then, calls for a dial east of the standard meridian with the dial slow, and our formula tells us that

$$\text{S.T.} = \text{L.A.T.} - L + e.$$

The L.A.T. is given as 10:25 A.M.; the longitude correction is 3°45′ or 15 minutes of time, and the equation of time on April 8 is $2^{\text{m}}06^{\text{s}}$ slow (see Appendix, Table A.1). Substituting these values in the appropriate equation, we get

$$\text{S.T.} = 10{:}25 - 15 + 2^{\text{m}}06^{\text{s}} = 10{:}12{:}06.$$

The correct Central Standard Time, then, is 10:12:06. Since we can seldom read a sundial to the nearest second, we may well round this off to 10:12 A.M.

Time When the Sun Souths. We often wish to know at what moment the sun will be on the meridian, due south of us (we commonly say of this situation that "the sun souths"). Since the sun souths at the moment of local apparent noon, we are merely asking what will be the standard time of local apparent noon. This involves nothing more than another application of the rules which we have just explained. For example, when will the sun south at Farmington, Connecticut, on June 16? With Farmington in the Eastern Standard Time zone with a longitude of 72° 49′, the longitude correction will be 2°11′ or 8^m44^s east of the standard meridian. On June 2 the equation of time is 0^m 23^s slow. Applying our rule, we get

$$\text{S.T.} = 12{:}00{:}00 - 8^m44^s + 0^m23^s = 11{:}51{:}39.$$

We know then that if our watch is correct the sun will be on the meridian directly south of us at 51 minutes and 39 seconds past 11 Eastern Standard Time.

Daylight Saving Time. In summer many localities use a time which is one hour ahead of their usual standard time—and many European communities use such advanced time throughout the year. If our clocks are keeping Daylight Saving Time we must remember to make allowance for the fact that they are just one hour faster than they would otherwise be. And if we want to design a sundial which will show Daylight Saving Time we merely compute the position of the hour lines in the ordinary way, but label them one hour off. That is, having computed the position of the hour line for 8 A.M., we label it 9 A.M., and having computed the position of the hour line for 3 P.M., we label it 4 P.M. Nothing could be simpler. Since most sundials, being outdoors, are used mainly in summer, it may be good judgment to label their hour lines in Daylight Saving Time.

3
The Noon Mark

In the days of the early settlers, before clocks were common, many a farm house had a *noon mark* near some southern window or on a southern porch, with the aid of which the housewife could watch the approach of noontime and know when to call the men in from the field for dinner. These old noon marks are often still visible in old houses, whose present tenants unhappily seldom recognize them or understand their purpose. It is also surprising that more people do not add such a simple, traditional, and useful feature to modern homes, especially since a simple noon mark can be laid out with only an hour's work at most.

In its simplest form a noon mark is merely a straight line on a level surface, along which the shadow of some vertical object falls at the moment of local apparent noon when the sun is on the meridian. Usually it was placed on the floor or window sill, where it caught the shadow of the side of the window, the jamb of a door, a post supporting the veranda roof, and so forth; but sometimes it appeared outside on a level lawn where it caught the noon shadow of a flagpole or other vertical object.

Finding the Meridian. The noon mark lies in the meridian. It runs due north and south, and to place it accurately we must have some means of finding a true north–south or meridian line. There are at least three ways of doing this.

(a) We may use a mariners' compass, in which case we must remember that in most places the compass needle does not point toward the true north, but veers more or less to the east or west of true north by an amount known as the *compass variation.* The amount of the variation changes slowly from year to year, but in general the compass needle points to the west of north in the eastern United States and to the east of north in the western part, with the magnetic compass giving true readings along a dividing line which runs from eastern Florida to eastern Wisconsin. If one wishes to find his meridian for sundial purposes by means of a magnetic compass, he should consult a local land surveyor or civil engineer to find out what correction he must make for the variation of the compass.

(b) The shadow of any vertical object will lie in the meridian at the moment of local apparent noon, when the sun souths. We have learned in Chapter 2 how to find when the sun will south, and if we hang a plumb line and mark the position of its shadow at the instant of local apparent noon we have our meridian safely marked.

(c) The ancients found the meridian, as astronomers still do, by the "method of equal altitudes." This involves marking the direction of the sun at some morning time, and marking it again in the afternoon when the sun has the same altitude. For example, we drive a nail perpendicularly into a board as at *A* in the upper part of Figure 3.1, and carefully level the board with the nail toward its southern end. We mark the position of the tip of the nail's shadow

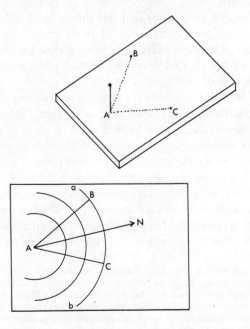

FIGURE 3.1 Finding the meridian by the method of equal altitudes.

at *B* in the morning, and again at *C* when the sun has the same altitude again that afternoon. To discover when the altitudes are equal, we draw several concentric circular arcs centered at *A*, the foot of the nail, as shown in the lower part of Figure 3.1, and when the tip of the shadow crosses one of these arcs in the morning, as at *B*, we mark the spot. In the afternoon, when the tip of the shadow crosses the same arc, as at *C*, we mark the point again. We draw lines *AB* and *AC*, and bisect angle *BAC* with the line *AN*, which is then part of the meridian with the direction from *A* to *N* being the direction of true north.[1]

While these alternative methods of finding the meridian may well be useful for other purposes as we place sundials in position, when laying out a noon mark it will usually be easiest merely to mark the position of the shadow of the window frame or door jamb at the moment of local apparent noon. This is all that is required to find our noon mark. Thereafter the shadow will always fall along the mark at the moment of local apparent noon, and we can convert this to standard time by the methods of the preceding chapter. It may be a surprise to discover that this simple mark will serve as an adequate check on the accuracy of one's grandfather clock or pocket watch. The noon mark may be scratched on the floor or a windowsill or a wall. Old noon marks were often outlined on the floor with a row of bright tacks.

Instead of noting the time by the position of a shadow, some people prefer to note the position of a small beam of light. Let us cut a piece of black paper or cardboard to fit a pane of glass in a southerly window. We punch a hole in the paper about a quarter inch in diameter, and fix the paper over the window pane. A small beam of sunlight will now be projected through the hole, throwing a small spot of sunlight on the floor. This sunbeam will move across the floor from west to east as the sun moves across the sky from east to west. Let us mark the position of this spot just at local apparent noon on a winter's day when the noontime sun is low, and again on a summer's

---

[1] Purists might wish to make a correction for the change in the sun's declination between the morning and afternoon observations. The effect will usually be negligible for our purposes, but if one wishes to be careful he will make his determination at or close to the times of the solstices when solar declination changes very slowly. In Figure 3.1 we show points on but one of the circular arcs. In practice the observer will probably want to mark the shadow's coincidence with two or three arcs, partly to help insure against the chance that the sun may be obscured by a cloud at a critical afternoon moment, and partly so that he may average his results.

day when the noontime sun is high in the sky. This will give us two spots on the floor some distance apart, and if we connect them with a straight line we will have a noon mark. Thereafter whenever the sunbeam crosses the line it will be the moment of local apparent noon. The sunbeam can be read more accurately than the shadow, and the accuracy can be increased further by "averaging." The beam of light which moves across the floor will have the shape of a small ellipse. As it moves, first the eastern and then the western edge of the ellipse will cross our noon mark. If we note carefully the times of these two passages and take the average, we should get a value which is correct within 20–30 seconds at most. Let us illustrate. Suppose we have an accurate noon mark at McPherson, Kansas, in longitude 97°41'. On September 28 the eastern edge of the sunspot crosses the noon mark when our watch registers 12:21:20 P.M., and the western edge crosses at 12:22:30.[2] The average of these two observations is 12:21:55 P.M. by our watch; but the methods of the preceding chapter tell us that the sun actually souths on this date in this latitude at 12:21:38 P.M., so our watch is 17 seconds fast.

Noon marks can be drawn just as easily on vertical surfaces, and in olden days castles, cathedrals, and other public buildings not uncommonly carried such vertical noon marks. Our frontispiece shows such a vertical mark, and others appear in Figures 3.2 and 3.3.

The Analemma. The simple noon mark denotes the moment of local apparent noon. If we use a projected sunbeam as our indicator it is quite possible to lay out a mark which will tell the moment of noon by standard time without any corrections. This will no longer be a simple straight line lying in the meridian, but will take the form of an elongated "figure eight." This queer shape arises from two factors: the changing declination of the sun and the changing values of the equation of time.

We can get some idea of the shape of the analemma by studying our tables of the sun's declination and of the equation of time. (See Tables A.1 and A.2 of the Appendix.) When the sun's declination is north in summer the high noon-day sun will cast its beam close to the south window; but when the sun's declination is south in winter the low noon-day sun will project its beam far back on

---

[2] If the aperture through which the sunbeam passes at the window is reasonably small, the projected sunspot will take just about one minute to cross the noon mark.

FIGURE 3.2 Noon mark on old French castle. From M. de la Prise,
Cadrans solaires (Caen, 1781).

FIGURE 3.3 Noon mark in a French garden. "Mid-day," by Nicolas Lancret, 1690–1743.

the floor toward the north. And at the same time that the noon sunbeam is moving north and south with the sun's declination, it moves east and west with the equation of time. For example, the sun is some 16 minutes "fast" early in November. It crosses the meridian 16 minutes before the clock shows noon, and when the clock does reach noon 16 minutes later, the sunspot has moved some distance east on the floor as the sun has moved toward the west. Whenever the sun is "fast" the noon sunspot lies east of the meridian; when the sun is "slow" the noon sunspot lies to the west. If each day at noon we drive a tack into the floor marking the position of the sunspot, we shall find as the days go by that the succession of tacks marks out an elongated figure eight on the floor—an analemma. Whenever the spot of sunlight crosses the analemma it is noon by watch time rather than by sundial time. If our watch was adjusted to Eastern Standard Time the sunspot will cross the analemma each day just at noon Eastern Standard Time;

and if the analemma has been carefully drawn it should be accurate enough to serve as a check on the accuracy of our clocks and watches.

One can start laying out an analemma by such noontime observations on any day of the year, driving a tack each day in the center of the sunspot just at noon by watch time. He will thus take a complete year for the task, but it will require no calculation whatever. If for a day or two the sun is obscured by clouds at noon, tacks can be interpolated by eye to fill the blank spaces. And if one marks the first day of each month by a double tack, the analemma will serve as a calendar as well as a clock. Whenever the sunspot crosses the analemma it is noon, and we can tell by the point where it crosses what day of the year it is. There may be some slight advantage in starting to lay out the analemma at noon on September 1 in a year following a leap year, as on September 1 in 1973 or 1977 or 1981, etc. This will center our observations between the calendar adjustments made each four years on "leap day," February 29th.

We should make sure, of course, that we have chosen a window which will catch the noonday sunlight all through the year. The sun may strike the window at noon during some seasons, but at other seasons the sunlight may be cut off by overhanging tree branches or by some other obstruction. If we persist patiently in our daily task of marking the sunspot we shall be surprised and pleased to see the tacks slowly trace out the elongated figure eight of the analemma until the finished diagram looks like that of Figure 3.4.

The actual dimensions of our analemma will depend on the height of the aperture above the floor and on the latitude. We should make sure in advance that we have enough clear floor space on the north side of the aperture to accommodate the entire analemma. The requisite distances are given in Table 3.1. The figures in this table must be multiplied by the height of the aperture above the floor. For example, if the aperture is 48 inches above the floor and the house is in latitude 40°, the analemma will run from a point (0.296) (48″) to a point (2.006) (48″), or from points 14.2 inches to 96.3 inches from the point on the floor vertically below the aperture.

Finding the Analemma by Computation. If one has the time, the training, and the patience he can compute mathematically the positions of the sunspots at noon on the various days, and then lay out the entire analemma by calculation. We start with three values: the latitude, ϕ, the sun's declination, D and the sun's hour angle, H.

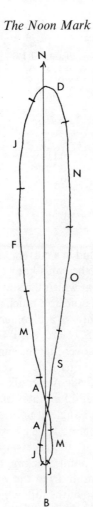

FIGURE 3.4 Approximate shape of an analemma as projected on a horizontal floor. Months are marked with their initial letters in sequence running counterclockwise around the larger loop and clockwise around the smaller loop of the "figure eight." The straight line, *BN*, is the meridian noon mark as it would appear if the analemma is constructed at a site exactly at a standard time meridian. At sites east of a standard time meridian, the figure eight would be displaced toward the right, while at sites west of a standard time meridian the figure eight would be displaced toward the left, in either case the straight line *BN* remaining where it is. When the sunspot crosses the line *BN* it is local apparent noon. When the sunspot crosses the part of the analemma which corresponds to the current date, it is noon by standard time.

TABLE 3.1
DIMENSIONS OF ANALEMMAS
IN VARIOUS LATITUDES

| latitude | distance from base | |
	shortest	longest
25°	0.026	1.130
30°	0.114	1.351
35°	0.203	1.632
40°	0.296	2.006
45°	0.394	2.539
50°	0.499	3.376
55°	0.613	4.915

The declination can be found in a current almanac or from Table A.2 of our Appendix. To find the sun's hour angle at noon on any day we find the standard time at which the sun souths, using methods of the preceding chapter, and note how much this differs from noon. We convert the difference to units of arc, remembering that each minute of time is equivalent to 15′ of arc, each 4 seconds of time to 1′ of arc. If, for example, we discover that the sun souths on a given day at 11:20 A.M. standard time, this differs from noon by 40 minutes. Reference to Table A.5 of our Appendix or use of the equivalents just listed will indicate that 40 minutes of time is the equivalent of 10° of arc, so the sun's hour angle at noon standard time is 10°.

If we are fortunate enough to have a copy of the relevant government tables[3] we can look up the necessary figures in them without further calculation at this point. Lacking these tables, we compute the sun's altitude and azimuth from the values of latitude, declination, and hour angle by use of the following formulas:

(1) $\tan M = \tan D/\cos H$
(2) $\tan a = (\cos M)(\tan H)/\sin(\phi - M)$
(3) $\tan A = \cos a/\tan(\phi - M)$

Here ϕ is the latitude, D is the solar declination, H is the sun's hour angle, M is merely an intermediate value computed for use in the subsequent formulas, a is the sun's azimuth at noon standard time

[3] *Tables of Computed Altitude and Azimuth*, Hydrographic Publication No. 214, available for about $4 to $5 through the Superintendent of Documents in Washington, D.C. One volume of these tables is published for each 10° of latitude, so you should be sure that you secure the right volume. For example, there is one volume for latitudes 30° to 39° inclusive, another for latitudes 40° to 49° inclusive.

and *A* is the sun's altitude at that same time. To illustrate, suppose
ϕ is 42°, *D* is 10° N (as in mid-April or late August) and *H* is the 10°
which we just found in our prior illustration. Substituting these
values in our three equations yields the following values:

$$M = 10°09' \qquad a = 18°12' \qquad A = 56°49'$$

These values for *a* and *A* would have been found without calculation
from the government tables listed in the footnote.

These results tell us that at the moment of standard time noon the
sun's altitude is 56°49', and instead of being now directly south (as it
would have been had the azimuth been zero) it bears 18°12' from the
south point. Since, furthermore, it is 11:20 A.M. local apparent time,
our common sense tells us that it must be 18°12' east of the meridian.
The sunspot on the floor will consequently lie 18°12' *west* of the
meridian.

If we have the meridian already marked on the floor (an ordinary
straight noon mark) it is now a simple matter to compute the
position of the sunspot. In Figure 3.5, *C* represents the aperture in the

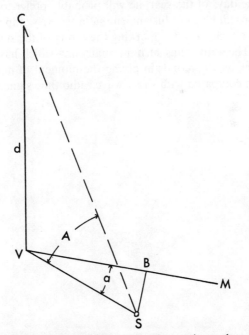

FIGURE 3.5 Computation of the position of one point on the analemma.

window; S is the sunspot on the floor; CS is the beam of light passing through the aperture; CV is a vertical line from the aperture to the floor (so d is the height of the aperture above the floor); M, B, V and S are points on the floor, which is assumed to be horizontal; VM is the meridian line or noon mark on the floor; angle VSC or angle A is the sun's altitude; angle BVS or angle a is the sun's azimuth or angle from the south; SB is perpendicular to VM; and by virtue of these constructions angles VBS, CVS and CVM are right angles. Given A and d, we have $VS = d \cot A$. Then $VB = VS \cos a$ and $BV = VS \cos a$ and $BS = BV \tan a$. Now we measure on the floor the distance VB along the noon mark and from B we lay off the perpendicular distance BS to the position of the sunspot at the required moment. It may be helpful to give the reminder that when the sun's declination is negative (from September 23 to March 20) the value of M will be negative, and this negative value must be subtracted algebraically from ϕ in our calculations.

When one remembers that it would be necessary to worry through these calculations 365 times to work out the analemma positions for each of the days of the year, he will probably prefer to pass up the calculations and lay out his analemma in the easy empirical way, by driving in a tack each day marking the center of the sunspot at that moment. The author has such an analemma which has never erred by as much as 10 seconds in giving the moment of noon standard time when compared with short-wave radio time signals.

4

The Equatorial Sundial

The equatorial sundial has the advantages that it can be made without mathematical calculation, and, unlike most other sundials, may be used in any latitude if properly set up. It gets it name from the fact that the dial plate lies parallel to the Equator. The *gnomon*[1] is a rod or pin set perpendicular to the dial plate in the center of the dial. This gnomon is parallel to the earth's axis and points toward the north celestial pole, close to the Pole Star. Figure 4.1 shows a side view of the dial in place. Angle *ACD* must be equal to the colatitude of the place where the dial is used.[2] Thus if the dial is set up at latitude 28°

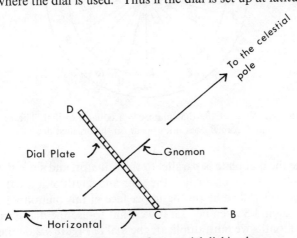

FIGURE 4.1 Side view of equatorial dial in place.

[1] The word "gnomon," accented on the first syllable and pronounced to rhyme with JOE-john, is derived from a Greek work meaning "one who knows." On many dials the gnomon is a triangular piece of metal or wood, one edge of which (called the *style*) casts the shadow which is used in telling the time. In equatorial dials a slim rod casts a line of shadow among the hour lines, and the time is read from the middle of this shadow.

[2] The colatitude of any place is the angle complementary to the latitude, found by subtracting the latitude from 90°.

angle *ACD* must be 62°. The dial can be used in any latitude merely by tilting the dial plate until angle *ACD* equals the colatitude.

Since the sun appears to move around the world's axis at the uniform rate of 15° each hour, the hour lines on the equatorial dial will be uniformly spaced 15° asunder. The dial plate will be laid out like that shown in Figure 4.2, with the line running from *B* to 12 lying in the meridian.

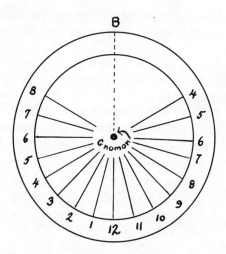

FIGURE 4.2 View of the dial plate of an equatorial dial. The broken line from *B* does not appear on the finished dial.

Since the dial plate is parallel to the Equator, and since the sun is north of the Equator for half the year and south of it for the other half, the sun will shine on the upper face of this dial only between March 20 and September 23. If we want to use the dial during the winter months we must duplicate the dial of Figure 4.2 on the lower face of the dial plate and extend the gnomon straight through the dial plate as it appears in Figure 4.1. The numbering on the under face must be reversed, running counterclockwise, but with the hour of noon still at the bottom. We need to include sufficient hour lines to run from the hour of earliest sunrise to the hour of latest sunset for our latitude. These limiting hours are tabulated in Table A.7 of our Appendix. Since the hours on the lower face will be used only in winter, they will run only from 6 A.M. to 6 P.M.

Showing Standard Time. The equatorial dial which we have described will show L.A.T., which can be converted to standard
time by methods described in Chapter 2. But this type of dial can be
so designed as to give readings in standard time directly without the
necessity of conversion. There are several ways of proceeding, but
perhaps the simplest is to construct the dial so that the slanting
gnomon is fixed but the dial plate is free to revolve around it, staying
always perpendicular to it. We attach a small auxiliary scale at the
edge of the dial plate as shown in Figure 4.3. This figure shows but a
small segment of the edge of the main dial plate bearing the hour
lines from 11 A.M. to 1 P.M. The spaces between these hour lines have
been subdivided into 5-minute intervals, equally spaced between the
hour lines. The auxiliary scale runs both to the right and the left of a
central zero mark with the spaces to the same scale as the main dial

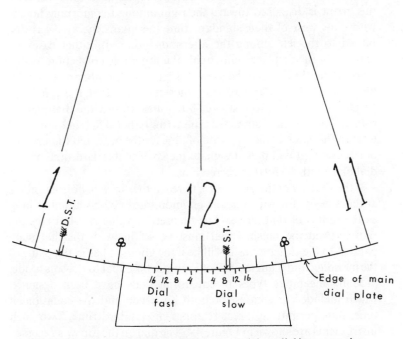

FIGURE 4.3 Detail of auxiliary scale on equatorial sundial incorporating
corrections for both the equation of time and the longitude, making it
possible to read either standard time or daylight saving time directly
from the sundial. The auxiliary scale is fixed and the main dial plate
rotates around the fixed gnomon.

plate. (Thus the distance corresponding to 2 minutes on the auxiliary scale should be the same as that which would denote 2 minutes on the main dial plate.) Since the equation of time never exceeds 16 minutes, we may limit the auxiliary scale accordingly. The zero line of the auxiliary scale lies in the meridian of the gnomon, and when the 12-o'clock line of the dial plate is set opposite the zero of the auxiliary scale, the dial will show L.A.T. But if the 12-o'clock line of the dial plate is turned opposite the value of the equation of time for the day we will automatically correct for the equation of time, and the dial will show local mean time directly.

If we wish also to make a longitude correction (as we should surely do if we are to build corrections into the dial at all) we merely mark an arrow on the main dial displaced from the 12-o'clock line by the amount of the longitude correction for our locality (see Chapter 2). If the dial is located east of its standard time meridian this arrow is displaced toward the right among the morning hours; with dials west of their standard time meridians the arrow is displaced to the left among the afternoon hours. In either case, the amount of the displacement equals the longitude correction for the place where the dial is to be used. In Figure 4.3 it is apparent that the dial is east of its standard time meridian and that the longitude correction amounts to just over 8 minutes. In use, the dial plate is rotated until the little standard time arrow (labeled S.T. in Figure 4.3) is set to the value of the equation of time for the date, and the shadow of the gnomon will now designate the correct standard time on the dial plate with no further corrections.

If the shadow of the gnomon is to reach the circle of hour numbers at the times of the solstices, the gnomon must be about half as long as the radius of that circle. (The theoretical value is 0.44 times the radius.) And the finished dial must be set up with the dial plate parallel to the Equator, as it will be if the gnomon lies in the plane of the meridian and if angle ACD of Figure 4.1 is equal to the colatitude.

Many ingenious types of equatorial dials have been designed which include corrections for both longitude and the equation of time, thus permitting direct readings in standard time. Two such instruments are shown in Figure 4.4. The dial on the top was designed by the Gryphon Corporation of Burbank, California. Its movable alidade, pivoted at its center in the bottom of the bowl, is rotated until the shadow of the notch in its upper end falls on the current date of the analemma at its lower end, as seen in the picture. The

.FIGURE 4.4 Two commercial sundials designed to give direct readings in standard time.

standard time is then read from the pointer on the upper end of the alidade. The instrument on the bottom in Figure 4.4 was artfully designed by Richard Schmoyer of Landisville, Pennsylvania. The gnomon on its axis is cut with a curiously curved slot through which the sun shines on the hour scale. The curve of the slot is engineered in such a fashion as to make allowance for the equation of time, and other adjustments are included for the latitude and for the longitude correction as well as for shifting to daylight saving time in season. Either of these dials is as accurate as most pocket watches.

5

The Horizontal Sundial

Horizontal sundials are far more common than all other types combined. Their popularity arises in part from the fact that they tell the time whenever the sun is shining, while many types of sundials can be used only during restricted hours of the day. In addition horizontal dials are comparatively easy to make and to set into place for use. They make beautiful and appropriate garden ornaments, and even a small garden can well find a spot for one at the end of a pathway or as an accent in a formal flower border. They may be made of a wide variety of materials ranging from the simple wooden dial turned out by a proud schoolboy to the magnificent silver or brass dials engraved and embellished by the finest artists and gracing the great museum collections. Figure 5.1 shows a beautiful and useful dial in the garden outside the Bruton Parish Church at Old Williamsburg, Virginia.

If a sundial is to tell the time accurately it must ordinarily be designed for the particular latitude in which it is to be used.[1] Unfortunately many of the commercial dials displayed in the stores are almost completely useless. Modern production methods turn them out by the hundred, all exactly alike, and modern merchandizing methods distribute them to all parts of the country (or the world) without regard to their suitability for the place where they are to be installed.

Morevoer, many of these mass-produced dials have not been designed correctly for any latitude whatever, but have merely been laid out by some draftsman who was familar with the general appearance of a sundial. It is not uncommon to see sundials displayed in a store with their gnomons attached back end to, so that the dial could not possibly indicate anything except the ignorance of the

---

[1] Later in this chapter we shall explain how to adjust a dial for use in a latitude other than that for which it was designed.

FIGURE 5.1 Ornamental horizontal sundial in a churchyard garden.

vendor. Yet a sundial of real reliability can be made easily, even by a twelve-year-old, and when it is properly designed for its location it will be both useful and ornamental.

General Appearance of the Horizontal Dial. The typical horizontal sundial consists of a flat horizontal piece, such as *ABCD* in Figure 5.2, called the *dial plate*. The hour lines are inscribed on the dial plate, radiating from the *dial center* at *O*. The 12-o'clock line, *OF*, lies in the meridian with *O* toward the south and *E* toward the north, and the gnomon is a roughly triangular plate, *OFE*, placed on *OF* perpendicular to the dial plate. The upper slanting line of the gnomon, *OE*, is called the *style*, and it is the shadow of the style, falling among the hour lines, which indicates the time of day. Angle *EOF* of the gnomon must be equal to the latitude of the place where the dial is to be used.

The Latitude. Before we can design a sundial we must know the latitude of the place where it is to be used. This latitude is ordinarily symbolized by the Greek letter ϕ (phi) and can be found from a large-scale map with sufficient accuracy. Particularly useful are the

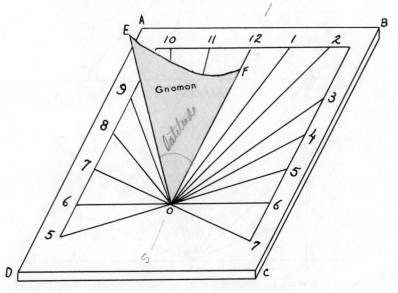

FIGURE 5.2 A typical horizontal sundial.

maps of the U.S. Geological Survey, from which the latitude can be read usually to the nearest minute or even closer. If you have trouble finding your latitude from maps, you should call for help from a local surveyor or civil engineer, or from a teacher of high school mathematics or geography. You should try to find the latitude at least to the nearest minute, keeping in mind that one minute of latitude corresponds to roughly 1.15 miles on the ground. While you are at it, it will be wise to find the longitude also, since we shall want to use it for other purposes later.

Laying Out the Hour Lines Graphically. There are several ways of proceeding to lay out a sundial. We shall here start with the simplest in that it requires no mathematical calculation whatever, and we shall illustrate the method with Figure 5.3. This working diagram is

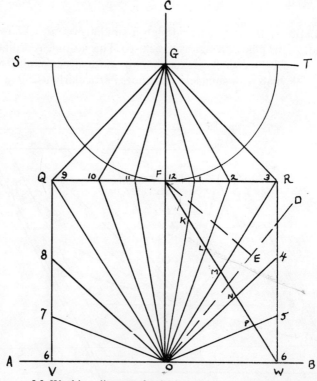

FIGURE 5.3 Working diagram for laying out a horizontal sundial by graphic methods.

drawn on a large piece of paper, larger than the finished dial, and when we are through we trace off from this working diagram just those lines which will appear on the finished dial. Our steps are as follows:[2]

(1) Draw *AB* and *CO* mutually perpendicular intersecting at *O*. In the final dial *AB* will be the 6-o'clock hour line and *CO* will be the 12-o'clock hour line.

(2) Draw *OD* making angle *COD* equal to the latitude. In our illustrative diagram *COD* is taken at 40°.

(3) At any point on *OD*, as at *E*, draw *EF* perpendicular to *OD* and intersecting *OC* at *F*. The length taken for *OE* will determine the size of the final working diagram, and can be taken longer or shorter as one wants a larger or a smaller diagram to work with.

(4) On *OC* lay off *FG* equal to *EF*.

(5) Through *G* and *F* draw *SGT* and *QFR* parallel to *AB*.

(6) Draw a semi-circular arc centered at *G* and based on *ST*. Any radius may be used. Here we took a radius equal to *GF*.

(7) From *GF* lay off along the semi-circular arc three equal arcs of 15° in each direction—15°, 30° and 45° each way from *GF*.

(8) From *G*, through the divisions of the semi-circular arc just found, draw straight lines intersecting *QR* at the points marked 9, 10, 11, 1, 2 and 3.

(9) From *O* draw straight lines to the six intersections just found on *QR*—that is, from *O* to 9, 10, 11, 1, 2 and 3. These will be the hour lines in the finished dial for 9 o'clock, 10 o'clock, 11 o'clock, etc.

(10) Draw *FW* parallel to the 9-o'clock hour line (that is, parallel to *QQ*) intersecting *AB* at *W*. This line *FW* will intersect the hour lines of 1, 2 and 3 at *K*, *L* and *M* respectively.

(11) From *M* lay off along *FW* the distances *MN* equal to *ML*, and *MP* equal to *MK*. The distance *MW* should be equal to *MF*.

(12) From *Q* and *R* draw *QV* and *RW* parallel to *CO*.

(13) From *O* draw lines through *N* and *P* intersecting *RW* at 4 and 5. These are the hour lines for 4 P.M. and 5 P.M.

(14) On *QV* make *Q8* equal to *R4* and *Q7* equal to *R5*.

(15) Draw straight lines from *O* to 7 and 8. These will be the hour lines for 7 A.M. and 8 A.M.

We now have the hour lines from 6 A.M. through 6 P.M. We should add hour lines before 6 A.M. and after 6 P.M. to include all hours from

---

[2] There are many possible graphic approaches. This one is based on that published in Paris in 1790 by Dom Francois Bedos de Celles.

the time of earliest sunrise to the time of latest sunset in our latitude. Our dial was designed for use in the latitude of 40°, and Table A.7 in our Appendix tells us that the earliest sunrise in that latitude is at 4:30 A.M. and the latest sunset falls at 7:33 P.M. We may, then, want to include hour lines from 4 A.M. to 8 P.M. inclusive. These added hour lines can be drawn on our diagram without difficulty, since any morning hour line is a straight-line extension of the corresponding afternoon hour line and vice versa. For example, if we extend the 5 P.M. hour line of Figure 5.3. from 5 straight through *O* and on down to the left this extension will be the hour line for 5 A.M.; and if we extend the line from 8 through *O* on down to the right we shall get the hour line for 8 P.M. Since we have already on our diagram the hour lines from 6 A.M. through 6 P.M., we can get any other hour lines which we wish merely by extending them through the dial center at *O*. The extended lines for 5 A.M. and for 7 P.M. appear in Figure 5.2.

Having the hour lines for our finished dial, we next lay out the gnomon. It will be a triangle similar to *FOE* of Figure 5.3. The angle *FOE* must be equal to the latitude. We set this gnomon up with its base on the 12-o'clock line, and perpendicular to the dial plate, as shown in Figure 5.2.

We say that the gnomon should be set up vertically on the 12-o'clock line, and this will be simple if the gnomon is made of thin material such as a plate of metal, If, however, our gnomon has appreciable thickness our diagram must be cut in half along line *OF* and the two halves separated by the thickness of the gnomon, as in Figure 5.4. Then one side of the upper slanting side of the gnomon will serve as the style in the morning and the other in the afternoon. Perhaps it should go without saying that after the diagram of Figure 5.3 is completed, the final hour lines will be traced off onto a clean piece of paper or directly onto the dial plate, omitting the other construction lines which appeared in the working diagram.

Laying Out the Hour Lines in a Unit Square. This method is even simpler and faster than the graphic method. We illustrate it with Figure 5.5.

(1) Lay out a square, *ABCD*, with the length of each side equal to 1.000 unit. Thus the side may be 1 foot long or 1 meter or 1 unit of any other size. If, for example, the side of the square is 15 inches, the

15 inches represent 1 unit, and all lengths given hereafter would be multiplied by 15 to reduce them to inches.

(2) Take the tangent values for the various hours from Table A.9 of our Appendix, using the line in that table which corresponds to the latitude in which the dial is to be used. We illustrate again with a dial for latitude 40°. If the desired latitude is not to be found in the table, one can interpolate between the values found there. Thus if our latitude is 38°20′, which lies one-third of the way from a latitude of 38° to a latitude of 39°, we may use values one-third of the way from those given in the table for 38° to those given for 39°.

(3) The Appendix table gives the following values for the hour lines of a horizontal sundial at latitude 40° :

11 A.M. and 1 P.M.	.1722
10 A.M. and 2 P.M.	.3711
9 A.M. and 3 P.M.	.7779v
8 A.M. and 4 P.M.	.4491v
7 A.M. and 5 P.M.	.2084v

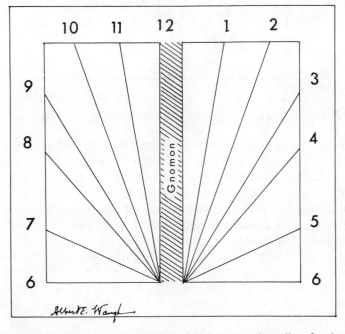

FIGURE 5.4 Hour lines of a horizontal dial separated to allow for the thickness of the gnomon.

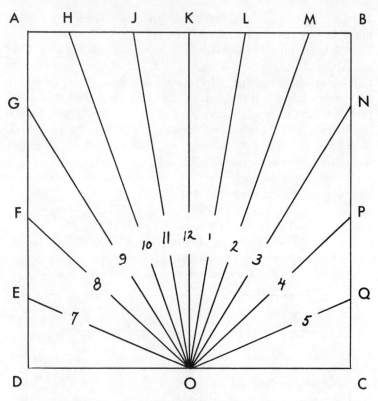

F IGURE 5.5 Horizontal sundial for latitude 40° laid out in a unit square.

(4) Locate *K* at the middle of *AB* and *O* at the middle of *CD*. A straight line from *O* to *K* will be the 12-o'clock hour line. The base of the square, *DC*, will be the 6-o'clock hour line.

(5) All measurements from the Appendix table are to be laid off along the top of the square from *K* toward *A* or toward *B unless* the tabulated measurement is followed by the letter "v" meaning "vertical." Measurements followed by "v" are to be laid off along the vertical sides of the square, starting at the bottom, and running upward from *C* toward *B* or from *D* toward *A*.

(6) Morning hours are shown on the left half of the square, being measured from *K* toward *A* or from *D* toward *A*. Afternoon hours are at the right, measured from *K* toward *B* or from *C* toward *B*.

(7) To find the hour line for 11 A.M. we start at *K* and measure 0.1722 units toward *A* to *J*. A straight line from *O* to *J* will be the

hour line for 11 A.M. To find the hour line for 4 P.M. we measure 0.4491 units vertically upward from C toward B to P, and a straight line from O to P is the hour line for 4 P.M.

(8) Had each side of our square been 15 inches long, we would have multiplied the tabulated figures by 15 to get results in inches. For example, point J should be 15 times 0.1722 inches from K, or 2.5830 inches, or about $2\frac{37}{64}$ inches from K. We always multiply the tabulated figures by the actual length of one side of our square to get results in the units in which we are working.

(9) The gnomon for this dial is made by the same rules which we have already explained for another case on page 40.

Laying Out the Hour Lines with Tabulated Angles. Using Table A.10 of our Appendix we may lay out the angles of a horizontal dial in short order without computation. Again we illustrate with a dial for latitude 40°, using the appropriate column of the Appendix table; and again, if our latitude lies between two of those shown in the table, we may interpolate to find values for our own latitude. We illustrate the process with Figure 5.6.

(1) Lay off lines AD and OK mutually perpendicular and intersecting at O. AD will be the 6-o'clock hour line and OK the 12-o'clock hour line.

(2) Table A.10 of our Appendix tells us that we want the following angles for the various hour lines at 40° latitude:

11 A.M. or 1 P.M.	9°46′
10 A.M. or 2 P.M.	20°22′
9 A.M. or 3 P.M.	32°44′
8 A.M. or 4 P.M.	48°04′
7 A.M. or 5 P.M.	67°22′

The table actually lists only the afternoon hours, but the angles for 2 P.M. and for 10 A.M. will be the same, since both times are equidistant from noon. Any morning hour will have the same angle as that afternoon hour which is the same distance from noon.

(3) Lay out the hour lines with the angles from OK just given. That is, make angle JOK and KOL equal to 9°46′ for 11 A.M. and 1 P.M.; make angles HOK and KOM equal to 20°22′ for 10 A.M. and 2 P.M., and so forth. Put morning hours on the left and afternoon hours on the right.

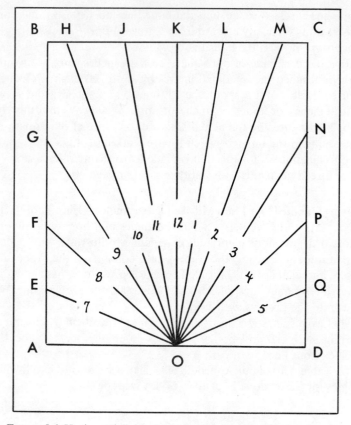

FIGURE 5.6 Horizontal dial for latitude 40° laid out by tabulated angles.

(4) Morning hour lines may be extended through *O* downward toward the right to get the hour line for 12 hours later; and afternoon hour lines may be extended through *O* downward to the left for the morning hour line 12 hours earlier.

(5) The gnomon is laid out by the same rules already explained on page 40.

(6) In the finished dial the numbers for the hours would often appear in the outer margin where the letters appear in Figure 5.6. In this figure the numerals were moved inward to make room for the letters which were used in our explanation but which would not, of course, appear in the finished dial.

Laying Out the Hour Lines by Computation. Instead of looking up the angles of the hour lines in our table, we can compute them mathematically for our precise latitude. For those with the requisite background the process is easy. For horizontal sundials we use either of the following formulas:

$$\tan D = (\tan t)(\sin \phi)$$
$$\log \tan D = \log \tan t + \log \sin \phi$$

In either of these forumlas D is the angle which the hour line makes with the 12-o'clock line, t is the time measured from noon in degrees and minutes of arc, and ϕ is the latitude of the place where the dial is to be used. In finding the value of t we take the number of hours and minutes before or after noon. Thus 2:35 P.M. and 9:25 A.M. are each 2 hours and 35 minutes from noon (usually symbolized in our work as 2^h35^m). But we have already learned that each hour of time is the equivalent of 15° of arc, and each minute of time to 15′ of arc. Using these equivalents (or Table A.5 of our Appendix) we see that 2^h35^m of time is the equivalent of 38°45′ of arc, and this is the value of t which we would use in our calculations if we wanted the hour line for either 2:35 P.M. or 9:25 A.M.

Let us assume that our sundial is to be used in Nashville, Tennessee, in latitude 36°10′. If we are to compute the angles for the hour lines and also for the intervening half hours we would make up a table like Table 5.1. In the first column we list the times of the hour lines for the hours and half hours. In the second column we reduce

TABLE 5.1
COMPUTATION OF ANGLES OF HOUR LINES

time	t	log tan$t$	log sinϕ	log tan$D$	D
11:30 or 12:30	7°30′	9.11943	9.77095	8.89038	4°27′
11:00 or 1:00	15°00′	9.42805	9.77095	9.19900	8°59′
10:30 or 1:30	22°30′	9.61722	9.77095	9.38817	13°44′
10:00 or 2:00	30°00′	9.76144	9.77095	9.53239	18°49′
9:30 or 2:30	37°30′	9.88498	9.77095	9.65593	24°22′
9:00 or 3:00	45°00′	0.00000	9.77095	9.77095	30°33′
8:30 or 3:30	52°30′	0.11502	9.77095	9.88597	37°34′
8:00 or 4:00	60°00′	0.23856	9.77095	0.00951	45°38′
7:30 or 4:30	67°30′	0.38278	9.77095	0.15373	54°56′
7:00 or 5:00	75°00′	0.57195	9.77095	0.34290	65°35′
6:30 or 5:30	82°30′	0.88057	9.77095	0.65152	77°25′

these to values of *t* by the method just explained. The third column contains the values of log tan *t* for each value at its left. The fourth column contains the value of log sin ϕ, which in our case is log sin 36°10′ or 9.77095. The values in the fifth column are sums of the two values at the left, and are values of log tan *D*. From these we get the values of *D* in the last column. These values of *D* are the angles which the various hour lines should make with the 12-o'clock line, and we proceed to chart them on a diagram exactly as we did when the angles were taken from a table on page 219. As one would expect, the values which we have just computed are very close to those which we would find in Table A.10 of the Appendix for the latitude 36°, since we have computed them for latitude 36°10′.

Longitude Correction for Horizontal Dials. The horizontal dials which we have so far described will record local apparent time, which, as we have seen, differs from clock time in two particulars: by the equation of time, and by the longitude correction. The equation of time varies from day to day, but the longitude correction is fixed for any given locality and we can make allowance for it in designing the dial. What we want is a dial which tells the local apparent time, not of our own meridian, but of our standard meridian. Thus we just designed a sun dial for Nashville, Tennessee, in latitude 36°10′. The longitude of Nashville is 86°50′ and clocks in Nashville are set to Central Standard Time, which is the time of the 90th meridian. Thus Nashville is 3°10′ east of its standard meridian. But 3°10′ is the equivalent of 12ᵐ40ˢ of time, and when it is, for example, 3:15 P.M. L.A.T. at Nashville, it is 12ᵐ40ˢ earlier at the standard meridian— or 3:02:20 P.M.

Let us make up a little table showing the L.A.T. at Nashville corresponding to selected L.A.T.'s at the standard meridian, the Nashville times being always 12ᵐ40ˢ later:

Local Apparent Time	
at 90° meridian	at Nashville
12 M	12:12:40 P.M.
1 P.M.	1:12:40 P.M.
2 P.M.	2:12:40 P.M.
3 P.M.	3:12:40 P.M.

If we now compute the hour lines for the times shown in the right-hand column but label them with the hours of the left-hand column, the dial will have the corrections for longitude built into it. One must still correct for the equation of time if he wants to compare with the clock, but it will no longer be necessary to make any correction for longitude. Of course, the hour lines will no longer be placed symmetrically. The gnomon will stand at the center of the dial running due north and south, but the hour line for 12 noon will now be drawn where the hour line for 12:12:40 P.M. L.A.T. would be on an ordinary dial, slightly to the right of the gnomon; the hour line for 3 P.M. would be drawn to the right of the gnomon in the position where the hour line for 3:12:40 P.M. would fall on an uncorrected dial. Longitude corrections may be incorporated in any other type of sundial in the same way. We find the amount of the longitude correction in hours and minutes, and add it to the hours to be shown on the dial if we are east of the standard meridian, or subtract it if we are west of the standard meridian. We then compute the hour lines for these adjusted times.

Adjustment to New Latitudes. We have emphasized the fact that a horizontal dial will perform accurately only in the latitude for which it was computed. This is the case *if* the dial is placed with the dial plate horizontal, as intended. If, however, we come into possession of a horizontal dial which was designed for some other latitude, we can still use it by making a minor adjustment.

First we must know the latitude for which the dial was originally designed. Preceding parts of this chapter should have alerted us to the fact that there are two clues which we can use in finding this latitude. First, the angle of the gnomon is equal to the latitude; and second, the angles which the various hour lines make with the 12-o'clock line depend on the latitude.

Let us illustrate with an actual case. Some years ago the author ran across in an antique shop a beautiful but simple engraved brass sundial carrying the date 1694. Careful measurement of the gnomon showed that the style makes an angle of about $52\frac{1}{2}°$ with the dial plate. The angles which the various hour lines make with the 12-o'clock line were then measured[3] with the following results:

hour line (t):	1 P.M.	2 P.M.	3 P.M.	4 P.M.	5 P.M.
angle (D):	12.0°	24.5°	38.5°	54.0°	71.5°

[3] How are such measurements made? The angle of the gnomon was measured directly with a protractor. Then a rubbing was made of the hour lines on the dial plate, and their angles were carefully measured.

We have used the formula

$$\log \tan D = \log \tan t + \log \sin \phi$$

to compute the dial angles when we know the latitude; but we can reverse it to find the latitude when we know the dial angles. The formula then becomes

$$\log \sin \phi = \log \tan D - \log \tan t.$$

Substituting the values of D which we measured on our antique dial and solving the formula for each hour line we get the following:

time (t):	15°	30°	45°	60°	75°
latitude (ϕ):	52.5°	52.1°	52.7°	52.6°	53.2°

While these results do not agree exactly, they fall pretty close together, and they are consistent in suggesting that the sundial was originally designed for a latitude of about $52\frac{1}{2}°$. This also agrees with our measurement of the angle of the gnomon. And when the antique dealer remarked that he bought the dial in Coventry, England (latitude 52°25′ N), the evidence, compelling before, becomes a certainty.

Having found that the dial was designed for latitude $52\frac{1}{2}°$, how can I use it in my garden at latitude 41°49′ N? If I set it on a pedestal with the dial plate horizontal, as was originally intended, the style, making an angle of $52\frac{1}{2}°$ with the dial plate, will not point toward the north celestial pole. The angle is too steep. But if I tip the south end of the dial plate up (thus tipping the style down) I can overcome that error. The difference between my latitude (41.8°) and the latitude for which the dial was designed (52.5°) is 10.7°; so if I tip the south end of the dial plate up until the dial plate makes an angle of 10.7° with the horizon (at the same time tipping the style down 10.7°) I will have the dial in a position to work properly. In fact, the dial plate will now be exactly parallel to the position that it once had when it was placed horizontally in England—parallel then to the English horizon.

Figure 5.7. should clarify the situation. In this diagram we are standing on the west side of the dials looking toward the east. The dial at the left was originally designed for a latitude greater than that of the place where it is now used (as in the case of the English dial used in my garden). The angle of the style was too great, so a wedge was placed under the dial plate to tilt it downward. The dial at the right of the same figure was designed for a lower latitude than

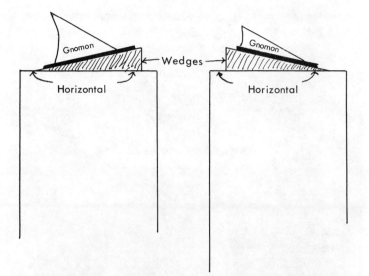

FIGURE 5.7 Two sundials adjusted for use in new latitudes. The left-hand dial has been tilted for use at a greater latitude than that for which it was designed; the right-hand dial for use at a lower latitude.

that where it is now used, as though I were using in my Connecticut garden a dial originally designed for Florida. In this case it would be necessary to increase the tilt of the style by inserting wedges under the north end of the dial plate, with the angle of the wedges again equal to the difference in the two latitudes. Figure 5.8 shows the actual English dial in the author's garden, viewed from the east side, with the wedge which alters the angle of the dial plate and the gnomon plainly visible.

Limiting Hour Lines. Since a horizontal sundial catches the sun's rays whenever the sun is above the horizon, we must show on the dial all hour lines from the time of earliest sunrise to the time of latest sunset. These limiting times are given for various latitudes in Table A.7 of our Appendix. Thus at Washington, D.C. (latitude 38°55′), the sun never rises earlier than about 4:34 A.M. nor sets later than about 7:30 P.M. Since it is customary to show the hour lines out to the next full hour, we would probably show hour lines from 4 A.M. to 8 P.M. for a Washington horizontal dial. At Anchorage, Alaska (latitude 61°10′) the earliest sunrise is about 2:20 A.M. and

FIGURE 5.8 Horizontal dial with the dial plate tipped out of the horizontal for use in a latitude other than that for which it was designed.

the latest sunset about 9:40 P.M., so a dial designed for use there would probably carry hour lines from 2 A.M. through 10 P.M. And beyond to the Arctic Circle, in the land of the midnight sun, we would have to include lines for all 24 hours.

6
Vertical Direct
South Dials

While most modern sundials appear as garden ornaments, with horizontal dials by far the most common, there was a time when sundials, like modern public clocks, appeared on the walls of public buildings or on a post in the public square to tell the time to all who passed. In such cases the dial plate was vertical, attached to the wall of the building, and the direction in which it faced depended on the orientation of the building. If the building is so oriented that its walls face exactly toward the cardinal points of the compass (north, south, east and west), dials attached to its walls are called *vertical direct dials*—vertical because the dial plate lies in a vertical plane, and direct because it faces directly toward one of the cardinal compass points. For example, the building shown in Figure 6.1 is exactly oriented with the major points of the compass (as contrasted with the building shown in Figure 10.1 on page 74). The wall shown at the bottom of Figure 6.1 runs in an east-west direction and faces due south. A dial placed on this wall would be called a *vertical direct south dial*. Note that vertical dials take their names from the direction their wall *faces*, which is at right angles to the direction the wall *runs*. In this chapter we shall deal with sundials placed on vertical walls which face directly toward the south.

Our task is simplified by the fact that the hour lines on any vertical direct south dial are precisely the same as those on a horizontal dial at the colatitude. In other words, to design a vertical direct south dial for latitude 50° we merely find the dial angles for a horizontal dial at latitude 40°.[1] But in the preceding chapter we learned several methods for finding the hour lines on a horizontal dial for latitude 40°; so our task is nearly finished before we start. Those same angles which we found before from tables or graphically or by computation can now be used for our vertical direct south dial.

[1] The colatitude is the remainder obtained by subtracting the latitude from 90°. Thus if our latitude is 50°, our colatitude is 90° − 50° = 40°.

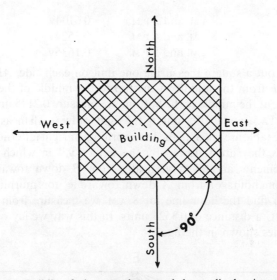

FIGURE 6.1 Building facing exactly toward the cardinal points of the compass. Vertical sundials attached to walls of this building will be vertical direct dials.

We say our task is "nearly" finished. There are, to be sure, a few new wrinkles. The numbering on a direct south vertical dial runs counterclockwise around the dial face, as contrasted with the clockwise sequence on a horizontal dial. On the vertical dial the dial center is at the top, with the 12-o'clock hour line running down vertically to the bottom, with the morning hours on the left and the afternoon hours on the right. And the gnomon makes an angle with the dial face equal to the colatitude rather than the latitude.

To eliminate possible confusion, let us find the hour lines for a vertical direct south dial in latitude 35°. We could use any of the methods of the preceding chapter. Let us try the method of the unit square. As with the horizontal dial, we lay out a square one unit long on each side, as in Figure 6.2. The values, from Table A.9 of our Appendix are taken from the line corresponding to the colatitude, 55°. (This table shows these colatitudes in the column at the right.) Using the appropriate row of the table we find the following distances for the various hour lines:

11 A.M. and 1 P.M.	0.2195
10 A.M. and 2 P.M.	0.4729

9 A.M. and 3 P.M.	0.6104v
8 A.M. and 4 P.M.	0.3524v
7 A.M. and 5 P.M.	0.1635v

We lay out a square measuring one unit on each side, *ABCD*. We draw *EF* from the middle of the top to the middle of the bottom. From *F*, at the middle of the bottom, we measure 0.2195 units to the left for 11 A.M. and 0.2195 units to the right for 1 P.M. All measurements from the table are taken to the left (morning) or right (afternoon) from *F unless* they are followed by the letter "v," in which case the measurements are taken vertically from *B* down toward *D* for afternoon hours or from *A* down toward *C* for morning hours. Thus to find the hour line for 8 A.M. we measure from *A* down toward *C* a distance of 0.3524 units. In this way we lay out all the hour lines shown in the figure.

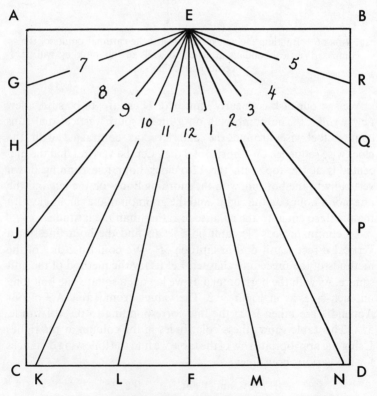

FIGURE 6.2 Vertical direct south dial laid out in a unit square.

On every vertical direct south dial the 6-o'clock hour line is a horizontal line at the top, and the 12-o'clock line is vertical. The sun can never shine on these dials earlier than 6 A.M. or later than 6 P.M., so we need not include any hour lines other than those shown in Figure 6.2. As with horizontal dials, if we use a gnomon of appreciable thickness our diagram must be cut in half along line *EF* and the two halves separated by the thickness of the gnomon. The gnomon must make an angle with the dial plate equal to the co-latitude (55° in our case). The base of the gnomon lies along line *EF*, with its point at the top so that the style (the side which casts the shadow) radiates from point *E* just as the hour lines do, although in a different plane. Figure 6.3 shows a vertical cross section of the dial in place against the wall.

From what has been said it should be clear that the vertical direct south dial and the horizontal dial are intimately related. The two are, in fact, sometimes combined. The combination may consist of

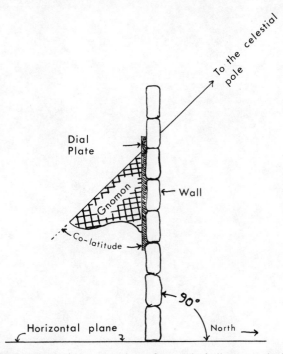

FIGURE 6.3 Vertical cross section of a vertical direct south dial in place on a wall.

two full-sized dials, or of two miniature dials which fold together to
form a "tablet dial" such as those described in detail in Chapter 16
and pictured in Figure 16.4. These combinations are easy to make,
are unusual enough to excite comment and, if carefully made, are
ornamental in use.

7
The Vertical Direct North Dial

Dials are seldom placed on walls which face directly toward the north, for obvious reasons. First, the sun is too far south in the sky to strike a north wall at all between September 23 and March 20, so a north dial is useless during the fall and winter months. But even during the spring and summer, the sun strikes a north wall only during early morning and late evening hours. During the middle of the day the north wall is shaded.

The maximum lighted period for a north-facing dial comes at the time of the summer solstice, and designing such a dial we should include only those hour lines which will be used on that date. The actual times depend on the latitude. A vertical direct north dial at latitude 40° north will receive the morning rays of the sun from the time of sunrise (which occurs at 4:30 A.M. in this latitude on this date) until 8:04 A.M. At this latter time the sun is due east in this latitude, and thereafter the sun is toward the south until 3:56 P.M., when the sun is due west. As the sun now falls into the northwest, the north dial again picks up its rays until sunset, which occurs at 7:33 P.M. on this date in this latitude. Thus the morning sun will strike the dial from 4:30 A.M. to 8:04 A.M., and the afternoon sun will strike it from 3:56 P.M. until 7:33 P.M. We would probably run these times out to the next whole hours, and show on our north dial the hour lines from 4:00 A.M. to 8:00 A.M. and the hour lines from 4:00 P.M. through 8:00 P.M., inclusive.

Table 7.1 shows the limiting times on vertical direct north dials at various latitudes in the temperate zones. In each case, the first morning time is the time of earliest sunrise and the last afternoon time is the time of latest sunset as shown in Table A.7 of the Appendix. The last morning time is the time when the sun is due east on June 21, which is the latest morning time that the sun's rays will ever reach the north dial; and the first afternoon time is the time

57

when the sun is due west on June 21, which is the earliest afternoon
time when the sun's rays will reach the dial. The figures in Table 7.1
should serve as a guide in designing a north-facing dial, but one will
usually carry the hour lines to the nearest or next whole hour.

TABLE 7.1
EARLIEST AND LATEST MORNING HOURS, AND EARLIEST AND LATEST
AFTERNOON HOURS, AT WHICH THE SUN STRIKES A DIRECT NORTH WALL IN
TEMPERATE LATITUDES

latitude	morning hours	afternoon hours
24°	5:12–11:08	12:52–6:52
26	5:07–10:11	1:49–6:56
28	5:02–9:38	2:22–7:00
30	4:58–9:15	2:45–7:05
32	4:53–8:56	3:04–7:09
34	4:48–8:40	3:20–7:15
36	4:43–8:26	3:34–7:21
38	4:38–8:15	3:45–7:27
40	4:30–8:04	3:56–7:33
42	4:23–7:55	4:05–7:40
44	4:17–7:47	4:13–7:47
46	4:09–7:39	4:21–7:55
48	4:00–7:32	4:28–8:04
50	3:50–7:25	4:35–8:13
52	3:39–7:19	4:41–8:24
54	3:27–7:14	4:46–8:36
56	3:13–7:08	4:52–8:51
58	2:56–7:03	4:57–9:07
60	2:35–6:58	5:02–9:28
62	2:07–6:53	5:07–9:53
64	1:29–6:49	5:11–10:31

The hour lines on a vertical direct north dial are merely continuations of the hour lines of the vertical direct south dial as seen from
the other side. We can imagine a vertical direct south dial drawn on a
thin translucent piece of paper. We hold it up, facing the south as
such a dial should be placed. If, now, we stand *behind* the dial plate
and look through it, we shall be looking at a vertical direct north dial.
If we were to extend the hour lines of the vertical direct south dial
through the dial center, and continue them on the other side, the
hour line which had represented 7 A.M. on the south dial would now
represent 7 P.M. on the north dial; the line which had represented
8 A.M. on the south dial would now represent 8 P.M. on the north dial,
and so forth. Morning hours on the south dial would be continued

as afternoon hours on the north dial; afternoon hours on the south dial would be continued as morning hours on the north dial.

This can perhaps best be visualized by means of Figure 7.1. The lower part of this figure is the vertical direct south dial of Figure 6.2, shown here reversed as it would be if we were to look at it from the back of the dial plate. It has been drawn in lightly, since it would not actually appear on the vertical direct north dial. At the top of the figure we have extended the hour lines as they would appear on the vertical direct north dial. We note from Table 7.1 that this dial, which was computed for latitude 46°50′, will be lighted by the sun from just after 4 A.M. until about 7:30 A.M.; and from about 4:30 P.M. until about 8:00 P.M. Consequently we should show the hour lines for 4:00 to 8:00 in the morning and from 4:00 to 8:00 in the evening.

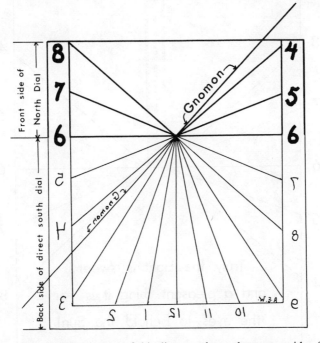

FIGURE 7.1 The lower part of this diagram shows the reverse side of a vertical direct south dial similar to that shown in Figure 6.2, as that dial would appear if viewed from the back by transmitted light. The upper part of this diagram is the corresponding vertical direct north dial for the same latitude. The hour lines on the upper north dial are continuations of the hour lines on the lower south dial.

Figure 7.2 shows the design of this vertical direct north dial, and Figure 7.3 shows a cross section of the dial as it would appear when in place on a north wall, with the gnomon making an angle equal to the colatitude of the place (43°10′, which we find by subtracting the latitude from 90°) with its point at the bottom. The base of the gnomon lies on a vertical line directly through the middle of the dial plate, and the 6-o'clock line is horizontal (unless we have introduced a longitude correction.)

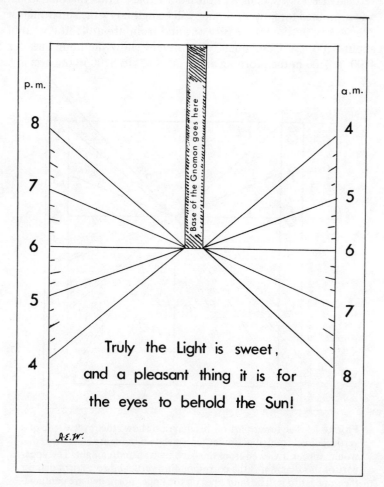

p.m.

8

7

6

5

4

a.m.

4

5

6

7

8

Base of the Gnomon goes here

Truly the Light is sweet,
and a pleasant thing it is for
the eyes to behold the Sun!

A.E.W.

FIGURE 7.2 Vertical direct north dial, latitude 46° 50′ north.

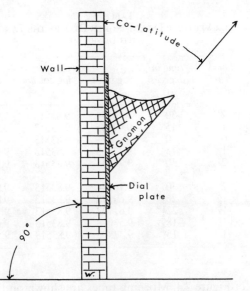

FIGURE 7.3 Cross section of a vertical direct north dial placed in its position against a north-facing wall.

Since the vertical direct north dial is merely a vertical direct south dial seen from the back, one can use any of the methods for making a vertical direct south dial as described in the preceding chapter and adapt them for the north face of the dial. The method of tangents can be somewhat simplified for these cases. Having decided which hour lines we wish to draw on the dial, we note how far each is (in time) from 6 A.M. or from 6 P.M. For example, if we are to show the hour lines for 4, 5, 6, 7 and 8 A.M., as we did in Figure 7.2, we show them as in Table 7.2, with 4 A.M. shown in the second column as 2 hours from 6 o'clock, and with 4:45 P.M. shown farther down in the column as $1\frac{1}{4}$ hours from 6 o'clock. In the third column these times are converted to degrees, with 15° equal to an hour; $7\frac{1}{2}$° equal to a half hour; $3\frac{3}{4}$° equal to a quarter hour; and so forth. The fourth column shows the logarithms of the tangents of the angles in the third column; and the fifth column shows the logarithm of the cosine of the latitude. We then use the formula

$$\log D = \log \tan t - \log \cos \phi,$$

where D is the distance which we are to measure on the tangent

TABLE 7.2
HOUR LINES FOR VERTICAL DIRECT NORTH DIAL BY THE TANGENT METHOD—
LATITUDE 46° 50′

time	hours from 6 o'clock	time in degrees (t)	log tan t	log cos φ	log D	D
4 A.M.	2	30°	9.76144	9.83513	9.92631	0.844
5 A.M.	1	15°	9.42805	9.83513	9.59292	0.392
6 A.M.	0	0°	——	9.83513	——	0.000
7 A.M.	1	15°	9.42805	9.83513	9.59292	0.392
8 A.M.	2	30°	9.76144	9.83513	9.92631	0.844
4 P.M.	2	30°	9.76144	9.83513	9.92631	0.844
4:15 P.M.	$1\frac{3}{4}$	$26\frac{1}{4}°$	9.69298	9.83513	9.85785	0.721
4:30 P.M.	$1\frac{1}{2}$	$22\frac{1}{2}°$	9.61722	9.83513	9.78209	0.606
4:45 P.M.	$1\frac{1}{4}$	$18\frac{3}{4}°$	9.53078	9.83513	9.69565	0.496
5:00 P.M.	1	15°	9.42805	9.83513	9.59292	0.392

diagram of Figure 7.4. Morning times are shown on the right-hand side of the diagram, with times before 6 A.M. measured up from the center and times after 6 A.M. measured down from the center. Afternoon times are shown on the left-hand side of the diagram, with times before 6 P.M. shown below the center and times after 6 P.M. shown above the center. The distances to measure up or down from the center are shown in the last column of Table 7.2, and are computed by means of the formula just given. For example, to compute the value for 4:15 P.M., we note first that 4:15 P.M. is $1\frac{3}{4}$ hours from six o'clock; and we convert $1\frac{3}{4}$ hours to $26\frac{1}{4}$ degrees of arc. From a standard table of trigonometric functions we discover that the logarithmic tangent of $26\frac{1}{4}°$ is 9.69298 and that the logarithmic cosine of the latitude, 46°50′, is 9.83513. Subtracting this latter value from the former we discover that the logarithm of D is 9.85785, and, taking the antilogarithm, we find that D is 0.721. This tells us to measure 0.721 units from the center of the diagram. Since it is an afternoon hour, we measure along the left-hand edge of the diagram, and since the time is before 6 P.M. we measure down from the center. The point in Figure 7.4 which represents 4:15 P.M. is, then, 0.721 units down from the center along the left-hand edge of the diagram. Table 7.2. is by no means complete, showing merely the full hours and certain illustrative quarter hours. The particular values which are computed in Table 7.2 have been transferred to Figure 7.4.

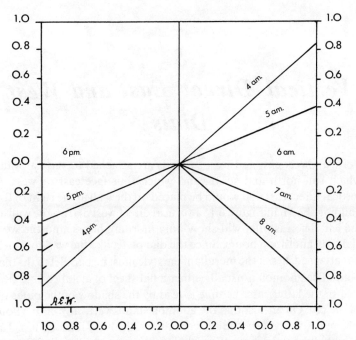

FIGURE 7.4 Laying out the vertical direct north dial by the tangent method, latitude 46° 50′. If each side of each of the four small squares is taken as 1.000 unit in length, the the 4-o'clock line runs up to a point 0.844 units above the center of the right side. Other values are shown in the last column of Table 7.2.

Finally, since the vertical direct north dial is like the vertical direct south dial as viewed from the back, we can note that the north dial:

(a) has the hours running clockwise, rather than counterclockwise.

(b) has the line for 12 o'clock midnight at the center of the top, rather than 12 o'clock noon at the center of the bottom.

8

Vertical Direct East and West Dials

Vertical direct east and west sundials are attached to vertical walls which run north and south, and which thus face east or west. A vertical direct east dial would be placed on the wall at the right of the building shown in Figure 6.1; a vertical direct west dial on the wall at the left. The east dial will show only morning hours, and the west dial only afternoon hours. Since the dial plate, like the wall to which it is attached, lies in the meridian, the style must be parallel to the dial plate. The gnomon is usually either a flat sheet of metal or a vertical pin. In the latter case the time is read by the shadow of the outer tip of the pin. On an east dial the gnomon stands vertically on the hour line of 6 A.M.; on a west dial vertically on the hour line of 6 P.M. The height of the style above the dial plate is always equal to the distance between the hour lines of 6 A.M. and 9 A.M. on a east dial, or 6 P.M. and 3 P.M. on a west dial. The hour lines are parallel with one another, so these dials have no dial center.

We can lay out these dials either graphically or by calculation. Our graphical method follows that of a famous book published in London in 1773,[1] and is illustrated in Figure 8.1. The steps are as follows:

(a) Draw *AB*, which, in the final dial, will be placed horizontally.

(b) At a convenient point, on *AB*, as at *O*, draw *COD* making angle *BOD* equal the latitude. In the figure, the latitude is 41°40′.

(c) Through *O* draw *EOF* perpendicular to *CD*. This line, called the equinoctial, will lie in the plane of the Equator, and line *COD* will lie parallel to the earth's axis.

(d) Draw a circle of any desired radius with *O* as center and divide the circumference into equal 15° arcs starting where *OC* intersects the circle.

(e) Draw *GH* and *JK* tangent to the circle and parallel to *EF*.

[1] Charles Leadbetter, *Mechanick Dialling* (London: Caslon, 1773).

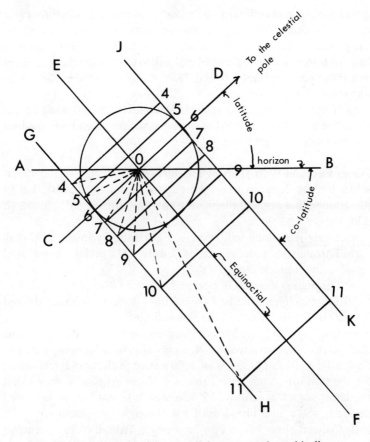

FIGURE 8.1 Vertical direct east dial constructed graphically.

(f) From *O* draw lines through the 15° divisions of the circle and extend them to intersect *GH* at 4, 5, 6, 7, 8, etc.

(g) Through these points 4, 5, 6, 7, 8, etc. draw lines parallel to *CD* and running from *GH* to *JK*. These are the hour lines for the morning hours of 4, 5, 6, 7, 8, etc. By noon, the sun will be shining parallel to the dial plate so the shadow will no longer fall on the dial. Hence we put hour lines on the east dial only from the time of earliest sunrise (see Appendix, Table A.7) until 11 or 11:30 A.M.

(h) The gnomon will have a height equal to the distance between the hour lines of 6 A.M. and 9 A.M., and will be set vertically on the hour line of 6 A.M.

(i) Line *AB* will ordinarily be transferred to the finished dial face very lightly, since its only purpose is to make sure that the dial has the proper slope. The dial must be attached to the vertical wall with line *AB* horizontal. The finished dial will usually show the hour lines and those parts of lines *GH* and *JK* running from the earliest hour line shown to the latest one.

(j) If the gnomon has a sensible thickness, the dial must be cut along the 6-o'clock line and the two parts separated enough to admit the gnomon between them.

Direct East and West Dials by Calculation. These dials lend themselves to easy design with a minimum of arithmetical work. Let us illustrate with a vertical direct west dial for latitude 35°, illustrated with Figure 8.2.

(a) Draw *AB* which will be placed horizontally in the final dial.

(b) Through any point on *AB*, as *O*, draw *COD* making angle *AOC* equal to the latitude (here 35°).

(c) Through *O* draw *EOF* perpendicular to *COD*.

(d) Draw *GH* parallel to *EOF* and at whatever distance desired from *EOF* for the width of the finished dial.

(e) Decide what hour lines to show on the finished dial. Since the sun will not shine on a western dial until shortly after noon, the hour lines will run from 12 : 30 P.M. or 1 : 00 P.M. to the hour of latest sunset for the latitude of the dial. Table A.7 of our Appendix shows that the latest sunset at latitude 35° is about 7 : 18 P.M., so we may well decide to show hour lines from 1 P.M. through 8 P.M. inclusive.

(f) Convert these hours into hour angles from 6 o'clock, remembering that each hour of time equals 15° of hour angle. Thus 5 P.M. and 7 P.M. lie 1 hour each side of 6 P.M. and hence convert to hour angles of 15°; 4 P.M. and 8 P.M. lie 2 hours from 6 P.M. and convert to 30°, and so forth. If one wishes to show the lines for half hours on the finished dial he can insert their hour angles at intervals of $7\frac{1}{2}°$.

(g) Make a table like Table 8.1, listing the hour lines, their corresponding hour angles and the natural tangents of those hour angles.

(h) The first three columns of Table 8.1. will serve as the basis for making any vertical direct east or west dial; but the last column is computed for our particular case, depending on the size of the finished dial which we desire. This size will depend on the dial's location. A dial high on a wall must be large, while one at eye level

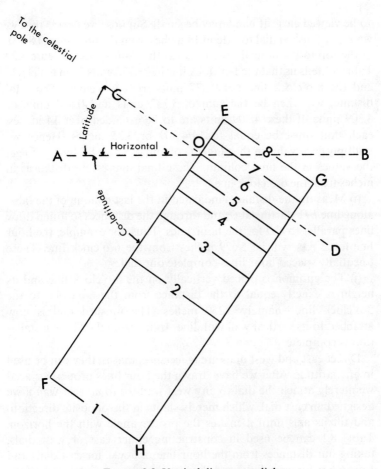

FIGURE 8.2 Vertical direct west dial.

TABLE 8.1
COMPUTING HOUR LINES FOR A VERTICAL DIRECT EAST OR WEST DIAL

time	hour angle	tangent of hour angle	inches on dial face
6	0°00′	0.000	0.000
5 & 7	15°00′	0.268	0.871
4 & 8	30°00′	0.577	1.875
3 & 9	45°00′	1.000	3.250
2 & 10	60°00′	1.732	5.629
1 & 11	75°00′	3.732	12.129

to be viewed close at hand may be small. Suppose we decide that we want our finished dial to extend 14 inches from the hour line of 8 P.M. at the top to the hour line of 1 P.M. at the bottom. (See Figure 8.2.) Table 8.1 tells us that the 1-o'clock line lies 3.732 units from 6 o'clock, and the 8-o'clock line lies 0.577 units from 6 o'clock. The total distance will then be the sum of 3.732 units and 0.577 units or 4.309 units. If these 4.309 units are to be extended over 14 inches, each unit must be equal to 14/4.309 or 3.25 inches. Hence we multiply each value in the tangent column of Table 8.1 by 3.25 to get the values in the final column. These final figures are distances in inches from the 6-o'clock line.

(i) Measure the distances indicated in the last column of the table along line *EF*, starting at *O*, and through the distances so found draw lines parallel to *CD* for the hour lines. Thus, for example, the hour line for 2 P.M. will lie 5.629 inches from the 6-o'clock line. These unequally spaced hour lines complete our dial.

(j) The gnomon is placed vertically on the 6-o'clock line and its height is exactly equal to the distance from the 6-o'clock to the 3-o'clock line—namely 3.250 inches. The finished dial is now attached to its vertical wall with line *AB* horizontal and the installation is complete.

Direct east and west dials are *universal*—that is, they can be used in any latitude. After we have drawn the hour lines properly spaced we merely attach the dial to any west wall (or to an east wall if we designed an east dial, which merely slants in the opposite direction) and tilt its axis until it makes the proper angle with the horizon. Table 8.1 can be used in constructing either east or west dials, taking our distances from the hour line of 6 A.M. for east dials and from 6 P.M. for west dials; and showing merely morning hours on the east dials and afternoon hours on the west dials. The direction of slant of the east and west dials follows that of Figure 8.1. for east dials and Figure 8.2. for west dials.

The gnomon of a vertical direct east or west dial always stands on the 6-o'clock line perpendicular to the dial face. It may take the form of a vertical pin just 1.000 unit long (that is, 3.250 inches long for the dial of Table 8.1, being always just as long as the distance from the 6-o'clock line to the 9-o'clock line on eastern and to the 3-o'clock line on western dials); or it may be a vertical plate of metal the upper edge of which, acting as the shadow-casting style, is parallel to the dial plate and 1 unit of distance above it. Thus a cross section of such a dial made vertically through the six-o'clock hour line would appear like Figure 8.3.

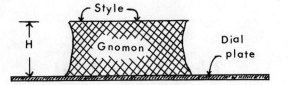

FIGURE 8.3 Cross section of east or west dial, showing gnomon in place. *H* is the height of the style.

9
The Polar Dial

Polar dials, like vertical direct east and west dials, are universal, being usable in any latitude. The hour lines lie parallel to one another and the gnomon stands vertically on the 12-o'clock hour line at the center with its upper edge (the style) parallel to the dial plate. The dial plate itself is not level, but lies parallel to the earth's axis. A horizontal line drawn across the dial face runs east and west and is perpendicular to the hour lines. The dial is symmetrical about the 12-o'clock line.

Graphic Construction. The steps in construction are similar to those used for the vertical east and west dials, and can be illustrated with Figure 9.1.

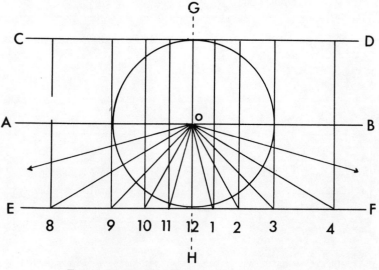

FIGURE 9.1 Graphical construction of a polar dial.

70

(a) Draw *AB* and at its center, *O*, draw perpendicular *GOH*.

(b) Draw *CD* and *EF* parallel to *AB* and equidistant from it at such distances that they give the desired width of the completed sundial.

(c) Draw a circle centered at *O* tangent to *CD* and *EF*.

(d) Starting at *OH* and running both ways divide the circumference of the circle into arcs of 15°, 30°, 45°, etc.; and from *O* draw lines through the points of division intersecting *EF*.

(e) At the points where the prolonged radii intersect *EF*, draw lines parallel to *GH* for the hour lines of the polar dial. These hour lines are numbered as shown in Figure 9.1.

Construction by Calculation. Lay out a rectangle of the desired size, and then proceed to draw the hour lines in it as follows:

(a) Vertically down the center draw the line *AB* as the 12-o'clock line. (See Figure 9.2.)

(b) Make up a table like Table 9.1 in which are listed, in successive columns, the hours to be shown on the dial, the hour angles of these hours from noon, and the natural tangents of these hour angles. Since the sun can never throw a shadow onto the dial plate of a polar dial until sometime after 6 A.M., and since the shadow leaves the

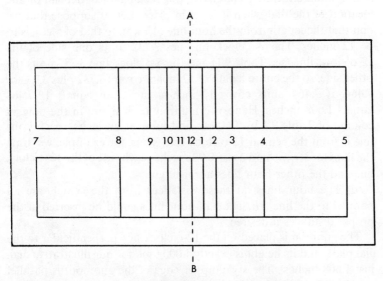

FIGURE 9.2 A polar sundial. The broken lines are used in construction but do not appear in the completed dial.

TABLE 9.1
COMPUTATIONS FOR CONSTRUCTION OF A POLAR SUNDIAL

hour line	hour angle	tangent of hour angle	distance on dial face
noon	0°	0.000	0.000
11½ & 12½	7½°	0.132	0.212
11 & 1	15°	0.268	0.431
10½ & 1½	22½°	0.414	0.666
10 & 2	30°	0.577	0.928
9½ & 2½	37½°	0.767	1.233
9 & 3	45°	1.000	1.608
8½ & 3½	52½°	1.303	2.083
8 & 4	60°	1.732	2.771
7½ & 4½	67½°	2.414	3.861
7 & 5	75°	3.732	5.971
6½ & 5½	82½°	7.596	12.145

dial sometime before 6 P.M., we usually take as our limiting hours 7 A.M. and 5 P.M. The table runs, however, from 6:30 A.M. through 5:30 P.M.

(c) Measure the length of the rectangle which you have laid out to accommodate the dial. The size of this rectangle will depend on the location of the dial when it is set in place. Let us suppose that we find that the length from the hour line of 7 A.M. to that of 5 P.M. is to be 12 inches. The 7-o'clock line lies 3.732 units one side of the 12-o'clock line (see Table 9.1) and the 5-o'clock line 3.732 units the other side, so the entire length of 12 inches covers 3.732 plus 3.732 or 7.464. If 7.464 units cover 12 inches, each unit equals 12/7.464 equals 1.608 inches. Hence we multiply each figure in the tangent column of Table 9.1 by 1.608 to find how far, in inches, each hour line is from the central 12-o'clock line. Thus, for example, we draw the lines for 10 A.M. and 2 P.M. 0.928 inches each side of the 12-o'clock line and the other hour lines accordingly.

(d) The hour lines, at these distances from the center, are all parallel to the line *AB*. The half-hour lines could be inserted at the proper distances if desired.

The gnomon is placed on the 12-o'clock line perpendicular to the dial plate, and its height is exactly 1.000 unit (for our illustrative dial, just 1.608 inches). The style (upper edge of the gnomon) is parallel to the dial plate. After the dial is completed it is set up for use with the 12-o'clock line and the gnomon both in the meridian and with the entire dial plate tipped up to make an angle with the horizon equal to the latitude, as seen in Figure 9.3.

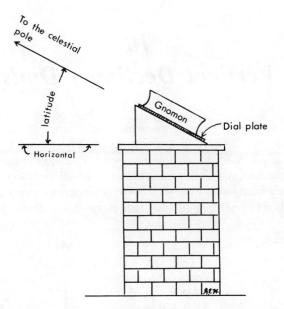

FIGURE 9.3 Cross section through a polar dial as seen from the west looking east.

10
Vertical Declining Dials

Vertical declining dials are those which are attached to vertical walls which do not directly face north, south, east, or west, but face instead some intermediate compass point. The walls of most buildings do decline, so one designing a dial for attachment to the wall of a building must usually design a "vertical decliner." Every declining wall falls into one or another of four categories as shown in Figure 10.1:

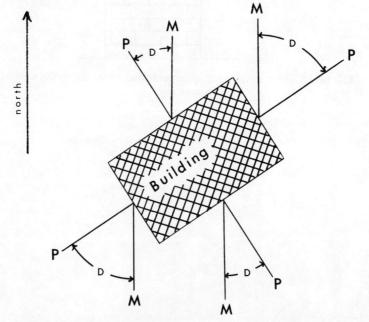

FIGURE 10.1 Measuring the declination of a wall. In each case, *P* is perpendicular to the wall and *M* lies in the meridian. The declination is measured by angle *D*.

(a) Southwest decliners, like the wall at the lower left of the figure.
(b) Southeast decliners, on the wall at the lower right.
(c) Northwest decliners, on the wall at the upper left.
(d) Northeast decliners, on the wall at the upper right.

We name the declining wall from the direction which we face if we stand with our back to the wall looking straight ahead at right angles to the wall. The amount of declination is measured by the angle between the meridian and a line at right angles to the face of the wall, as shown for each wall in Figure 10.1.

Before we can design a sundial to attach to such a declining wall we must know the latitude (as with other dials) and also the amount of the wall's declination. We shall describe a little later in this chapter the method of finding the wall's declination, but in our preliminary descriptions we shall assume that both the latitude (ϕ) and the wall's declination (D) are known. The amount of the declination is usually expressed as an angle accompanied by letters to show the direction which the wall faces. Thus in Figure 10.2 we have a wall, AB, with line OS lying in the meridian with S at its south end. Line OC is perpendicular to the wall. If measurement discloses that angle COS is 24° we would say that this wall declines S 24° E, which means that it faces somewhat toward the south, but has been twisted around toward the east through an angle of 24°

Graphical Method. Let us assume that we wish to design a sundial to attach to the vertical wall shown in Figure 10.2, and that the

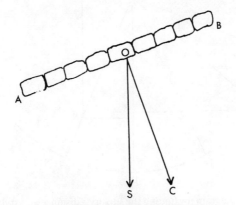

FIGURE 10.2 Declination of a wall.

building is located at latitude 40° N. We would usually abbreviate this by saying that our dial is at $\phi = 40°$ N and D is S 24° E.

Our first step is to find the *sub-style line*, which is the line on the dial plate upon which the gnomon is to be erected perpendicular to the dial plate. We have become accustomed to placing the gnomon on the 12-o'clock line, but with vertical decliners this is no longer the case. In all vertical dials the 12-o'clock line is vertical, but with vertical declining dials the line on which the gnomon is placed (called the *sub-style*) is twisted out of the vertical, lying to the right or east of the 12-o'clock vertical line and thus among the afternoon hour lines if the dial declines toward the west of south, but to the left or west among the morning hour lines with southeast decliners. The angle between the vertical 12-o'clock line and the sub-style is called the *sub-style distance*, or SD. To find this sub-style distance graphically we proceed as follows, using Figure 10.3.

(1) Draw *AB* as the horizontal line.

(2) At *C*, near the center of *AB*, draw *CD* perpendicular to *AB*.

(3) With any convenient radius draw the semi-circle *ADB* centered at *C*.

(4) Draw *CG* making angle *DCG* equal to the colatitude, here 50°. Since our dial declines toward the east, we place *CG* to the left of *CD*. For west decliners it would lie to the right.

(5) From *G* draw *GH* parallel to *AB*, cutting *CD* at *H*.

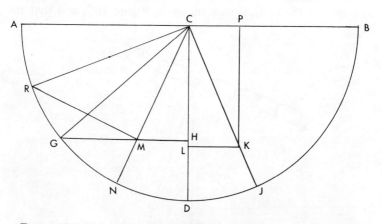

FIGURE 10.3 Diagram for finding the sub-style distance and style height of a vertical declining dial.

(6) On the side of *CD* opposite *CG* draw *CJ* with angle *DCJ* equal to the dial's declination (here 24°).

(7) On *CJ* lay off *CK* equal to *GH*.

(8) Draw *KL* parallel to *AB* intersecting *CD* at *L*.

(9) On *HG* lay off *HM* equal to *KL*.

(10) Draw *CMN*. This is the sub-style line, and angle *DCN* is the sub-style distance, SD. This ends our first step.

Having found the sub-style, on which the gnomon will be placed, we next find the *style height*, SH, which is the angle which the style makes with the dial plate. We continue illustrating with Figure 10.3, as follows:

(1) Draw *KP* parallel to *CD*.

(2) Find point *R* on the semi-circle so that *MR = KP*.

(3) Draw *CR*. This represents the style, and angle *NCR* is the required height of the style, SH.

Now that we have SD and SH we proceed to lay out the hour lines. This we could do by adding lines to Figure 10.3, but to avoid confusion we continue with a new diagram, Figure 10.4. In Figure 10.4, *AB* is horizontal and *CD* is vertical. *CN* is the sub-style, with angle *DCN* equal to angle *DCN* of Figure 10.3. And *CR* is the style, with angle *DCR* of Figure 10.4 equal to angle *DCR* of Figure 10.3. In other words, we have transferred essential lines from Figure 10.3 to our new diagram. Now in Figure 10.4:

(1) At any convenient point on *CN*, as *M*, draw *SMT* perpendicular to *CN*. The distance taken for *CM* will determine the scale of the final diagram.

(2) Draw *ME* perpendicular to *CR*.

(3) On *CN* lay off *MO* equal to *ME*.

(4) With *O* as center draw a circle of any desired radius. In Figure 10.4 the radius has been taken equal to *OM*.

(5) Call the intersection of *ST* and *CD* "*d*." Draw *Od*.

(6) Divide the circle into 15° arcs starting from *Od*.

(7) Draw lines from *O* through the 15° divisions on the circle and continue these radii until they intersect *ST* at *b, c, e, f, g, h* and *i*.

(8) Draw lines from *C* to the points just found on *ST*. That is, draw *Cb, Cc, Ce*, etc. These are the hour lines. They are numbered counterclockwise with *CdD* the 12-o'clock line.

These hour lines are now transferred from our work sheet to the dial plate, or perhaps directly to the wall itself. We will need the

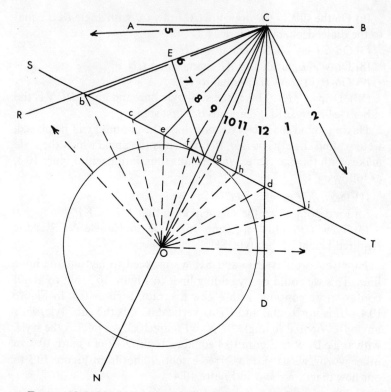

FIGURE 10.4 Graphic determination of the hour lines of a vertical declining dial. Lines *AB*, *CD*, *CN*, and *CR* are taken from Figure 10.3.

sub-style line and the hour lines, remembering that the 12-o'clock line must be vertical when the dial is in place. The gnomon goes on the sub-style line perpendicular to the dial plate.

The Computational Method. Vertical declining dials are easily laid out mathematically by those with the requisite background. Our problem falls into five steps:

  (a) Finding the sub-style distance, SD.
  (b) Finding the style height, SH.
  (c) Finding the difference in longitude, DL.
  (d) Finding the angle, AV, between the vertical 12-o'clock line and the hour line for 6 o'clock.
  (e) Finding the positions of the hour lines.

When we know the values of the latitude, ϕ, and of the declination of the wall, D, the first four of these steps involve the solution of the four following equations:

(1)	$\log \tan SD = \log \sin D + \log \cot \phi$
(2)	$\log \sin SH = \log \cos D + \log \cos \phi$
(3)	$\log \cot DL = \log \cot D + \log \sin \phi$
(4)	$\log \cot AV = \log \sin D + \log \tan \phi$

Since each of these equations is solved in terms of ϕ and D we can speed our work by organizing it in a table like Table 10.1.

TABLE 10.1
COMPUTING BASIC VALUES FOR A VERTICAL DIAL DECLINING S 24° E IN LATITUDE 40° N

D (24°)	log sin 9.60931	log cos 9.96073	log cot 0.35142	log sin 9.60931
ϕ (40°)	log cot 0.07619	log cos 9.88425	log sin 9.80807	log tan 9.92381
sums	log tan 9.68550	log sin 9.84498	log cot 0.15949	log cot 9.53312
	SD = 25°51.7′	SH = 44°24.7′	DL = 34°42.5′	AV = 71°09.4′

The sum of the first column gives us log tan SD, that of the second column gives log sin SH, and so forth—thus giving the sums necessary to satisfy the formulas; and from these sums we compute the first four of our required values, which, for our illustrative case, turn out to be: SD = 25°51.7′; SH = 44°24.7′; DL = 34°42.5′ and AV = 71°09.4′. We might well note before we leave the table that it contains a partial check on the accuracy of our arithmetic. If we have not erred, the sum which appears in the second column should be equal to the sum of the two sums in the first and third columns. In our case, then, 9.84498 should equal the sum of 9.68550 and 0.15949. This equality does in fact exist save for a discrepancy of one unit in the last decimal place, and we can assume that this reflects merely rounding off in the trigonometric tables. In our case we take the check as being satisfactory.

Now that we have four of our five basic values we proceed to compute the positions of the hour lines, using Table 10.2.

The figures of this table are derived as follows:

(a) Convert the value of DL (here 34°42.5′) to units of time (see Appendix, Table A.4), giving here a value of 2 hours 18 minutes 50 seconds. This tells us that the sub-style will lie at a time 2 hours 18 minutes 50 seconds from noon.

TABLE 10.2
COMPUTATION OF HOUR LINES OF VERTICAL DIAL DECLINING S 24° E IN
LATITUDE 40°. DIFFERENCE IN LONGITUDE IS 34° 42.5′

time	P	log tan P	log sin SH	log tan A	A
4 A.M.	85°17.5′	1.08427	9.84498	0.92925	83°17.3′
5 A.M.	70°17.5′	0.44586	9.84498	0.29084	62°53.6′
6 A.M.	55°17.5′	0.15948	9.84498	0.00446	45°17.6′
7 A.M.	40°17.5′	9.92830	9.84498	9.77328	30°40.9′
8 A.M.	25°17.5′	9.67442	9.84498	9.51940	18°17.9′
9 A.M.	10°17.5′	9.25907	9.84498	9.10405	7°14.5′
The sub-style falls here between 9 A.M. and 10 A.M.					
10 A.M.	4°42.5′	8.91573	9.84498	8.76071	3°17.9′
11 A.M.	19°42.5′	9.55414	9.84498	9.39912	14°04.4′
NOON	34°42.5′*	9.84052	9.84498	9.68550	25°51.7′†
1 P.M.	49°42.5′	0.07170	9.84498	9.91668	39°32.2′
2 P.M.	64°42.5′	0.32558	9.84498	0.17056	55°58.3′
3 P.M.	79°42.5′	0.74093	9.84498	0.58591	75°27.3′

*This value for noon always equals the difference in longitude.
†This value for noon always equals the sub-style distance.

(b) On east decliners the sub-style falls before noon; on west decliners after noon. Since ours is an east decliner the sub-style will fall 2 hours 18 minutes 50 seconds *before* noon at 9:41:10 A.M. (Had this been a west decliner the sub-style would have fallen at 2:18:50 P.M.) Since our sub-style will fall between 9 A.M. and 10 A.M. we note that fact by a line across Table 10.2 between these two hours.

(c) The figures in the first column of Table 10.2 are the hours which we will show on the finished dial.

(d) The second column lists what Holwell calls the "polar angles,"[1] which we represent by "*P*." We find their values thus: Opposite the hour of noon in this second column we enter the value of the difference in longitude as found in Table 10.1 (34°42.5′ in this case). As we move up and down the second column we add 15° for each hour moved away from the sub-style and subtract 15° for each hour moved toward the sub-style.[2] When we reach 10 A.M. by this method we get the value 4°42.5′, and if we are to move another 15° for 10 o'clock we will use up the 4°42.5′ and 10°17.5′ more. In other words to get to 9 A.M. we will move right past the sub-style and have

[1] John Holwell, *Clavis Horologiae* (London: 1712).

[2] If we are showing lines for the half or quarter hours on our dial, we add or subtract $7\frac{1}{2}°$ per half hour or $3\frac{3}{4}°$ per quarter hour.

10°17.5′ left over, which is the value entered opposite 9 A.M. in the second column. Now as we go on to 8 A.M., noon, and so forth, we are moving away from the sub-style, so we follow our rule of adding 15° per hour.

(e) We now make use of the formula for finding the angles of the hour lines on a vertical declining dial, namely

$$\log \tan A = \log \sin \text{SH} + \log \tan P,$$

where A is the angle which each hour line makes with the sub-style, SH is the style height taken from Table 10.1 and P is the polar angle of the particular hour line in Table 10.2. The remaining figures of Table 10.2 are merely those necessary to solve this formula.

(f) Each figure in the third column is the log tan of the corresponding figure in the second column.

(g) The figures in the fourth column, all identical, are the values of log sin SH taken from Table 10.1.

(h) The figures in the fifth column are sums of the two figures immediately at their left, and these sums give us log tan A.

(i) The figures in the last column are the angles, A, which correspond to the values of log tan A in the preceding column. They are the angles which the hour lines listed in the first column make with the sub-style. Here we can check again on the accuracy of our computations in two ways:

(1) The value of A for noon should be equal to the value of SD found in Table 10.1.

(2) If we subtract our value of SD from the value of AV (both in Table 10.1) the difference should be equal to the dial angle of the 6 A.M. hour line of Table 10.2. Here AV (71°09.4′) − SD (25°51.7′) = 45°17.7′ which differs from the 6 A.M. value of A in the last column of Table 10.2 by but a tenth of a degree, and we can assume that the checks are satisfactory.

Figure 10.5 shows the actual layout of this dial. We start with the horizontal line, AB, and the vertical line, CD. The vertical line represents the hour line of 12-o'clock noon. We lay off the sub-style line, CE, at the left of CD making angle DCE equal SD or 25°51.8′.[3] We then lay off the other hour lines, making the angles with the sub-style line called for in the last column of Table 10.2.

[3] Had this been a west decliner CE would have been on the right of CD by the value of SD.

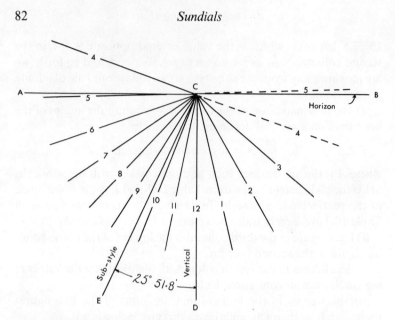

FIGURE 10.5 Hour lines for vertical dial declining S 24° E in latitude 40° N, laid out from the angles of Table 10.2.

We shall discuss later which hour lines should be shown on a vertical decliner (see page 83); but here let us remark that if we wish to include hour lines not included in Table 10.2 we can prolong the hour line of the hour which is 12 hours earlier or later. Thus the hour line for 4 P.M., not shown in the table, is merely a prolongation of the line for 4 A.M. Such prolonged lines for 4 P.M. and 5 P.M. are shown by broken lines in Figure 10.5.

Had our dial in latitude 40° N been for a wall which declined 24° *west* of south (all factors the same as before save that we have a west decliner rather than an east decliner), the computations of Table 10.1 would have been identical. But the sub-style would have fallen at 2:18:50 P.M. We would have entered the value of the difference in longitude (34°42.5′) as the value of P at noon, just as before, but now as we moved into the early afternoon hours we would have been moving toward the sub-style, and our values of P would have become smaller. As we moved from noon into the morning we would have been moving away from the sub-style, and our values of P would have become larger. And in Figure 10.5 the sub-style line, *CE*, would have been drawn 25°51.8′ to the right of *CD* rather than to the left.

Having laid out the hour lines, we next need to cut out the gnomon. This is usually a triangular piece, shaped something like that shown in Figure 10.6, with the style (the upper edge which casts the shadow) making an angle with the dial plate equal to SH, which in this case is 44°24.7′. The gnomon is set on the sub-style line of Figure 10.5 with its pointed end (the left in Figure 10.6) at the dial center, which is point *C* in Figure 10.5.

What Hour Lines to Include. No vertical dial can ever catch the sun's rays for more than 12 of the 24 hours in a day. A vertical direct south dial shows the hours from 6 A.M. to 6 P.M., but even if the sun rises well before 6 A.M. and sets long after 6 P.M. it does us no good to include hour lines for these early and late hours. If the sun is up then, it will lie north of the dial and will not shine on its face.

If we start to twist a vertical direct south dial toward the east, continuing to keep it vertical, we will begin to catch the sun before 6 A.M., but only at the cost of losing the sun before 6 P.M. If we twist it enough to catch the 5 A.M. sun, we will not catch the sun after 5 P.M. Again we are limited to a 12-hour span. And how do we know *which* 12 hours? There are several complicated mathematical rules, but simplest of all is the statement, "Include only those hour lines which lie below line *AB* in Figure 10.5." Reference to that diagram will demonstrate that, for the dial which we had there computed, the hour line for 4 A.M. lies above line *AB*, as does the hour line for 5 P.M. If we are going to limit ourselves to whole hours (not showing half hours or quarter hours) our earliest hour line should be that of 5 A.M. and the last one should be for 4 P.M. Since the 5 P.M. hour line lies so close to *AB* we might decide to include it even if the sun will never quite reach it.

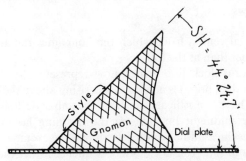

FIGURE 10.6 The gnomon of the dial computed in Tables 10.1 and 10.2.

Vertical North Decliners. One designing a sundial naturally chooses a sunny south wall; but there may be occasions when one wants a dial on a wall with northerly exposure. In Chapter 7 we learned how to design dials for walls facing directly north. Now we turn to walls which have a northerly aspect, but which decline somewhat to the east or west, like the two upper walls shown in Figure 10.1. Our procedure parallels for the most part that explained earlier in this chapter for south decliners.

While we did not explain it at the time, whenever one carries out the computations for any one vertical declining dial he finds at the same time all the angles of three other related dials. For example, the data of Table 10.1 can be used for any of the following dials in latitude 40°:

(a) A vertical south dial declining 24° east.
(b) A vertical south dial declining 24° west.
(c) A vertical north dial declining 24° east.
(d) A vertical north dial declining 24° west.

Let us suppose we want a dial to go on a wall facing N 24° W. Table 10.1 gives data for a dial facing S 24° E. These two walls are opposites—back to back. Since the latitude (40°) and the declination (24°) are exactly the same as before, Table 10.1 will be identical for both dials. We will find exactly the same values for SD, SH, DL, and AV. And in Table 10.2, the angles of the various hour lines with the sub-style line will be the same. In other words, all our computations have already been made. The northerly dial differs from its southerly counterpart, not in the mathematics, but in the position of the gnomon and the order of the hour lines. Our rules for north decliners are as follows:

(a) The dial center, from which the hour lines radiate, is at the bottom rather than at the top.
(b) The 12-o'clock line (which now represents 12-o'clock midnight rather than noon) is a vertical line rising above the dial center.
(c) If we have a northeast decliner, the sub-style line lies to the right of the midnight line among the morning hours; but with northwest decliners it lies to the left among the evening hours.
(d) The hour lines run in clockwise sequence rather than counterclockwise.

These rules will be more readily understood by reference to Figure 10.7. We start with the horizontal line *AB* and the vertical line *CD*. The sub-style distance is again 25°51.8′ and, with a west decliner, it lies to the left of *CD* among the evening hours. While the sub-style of our southeast decliner lay between the hour lines of 9 A.M. and 10 A.M., those of our northwest decliner lies between the lines of 9 P.M. and 10 P.M. Since the hours run in clockwise sequence we get the arrangement of Figure 10.7, using the angles for the hour lines which we found in Table 10.2.

Limiting Hours on North Decliners. A table such as Table 10.2 gives us the positions of twelve hour lines, and by prolonging them we can get the positions of all 24 hour lines of the day. But there is no sense in showing 24 hour lines on a dial which can catch the sun's rays for but a few hours in the morning and a few more in the evening. Table A.7 of our Appendix tells us that the earliest sunrise at latitude 40° comes at 4:30 A.M. and the latest sunset at 7:33 P.M. These immediately set limits. But inspection of Figure 10.7 shows that the hour line for 5 A.M. is barely below the horizon, and there will be no

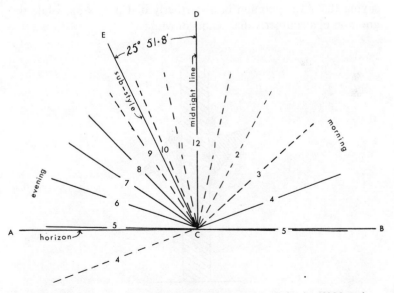

FIGURE 10.7 Work sheet for laying out a dial in latitude 40° N and declining N 24° W.

use in showing later morning hour lines than that. And on the left-hand side of Figure 10.7 we note that the afternoon hour lines are below the horizon until just before 5 P.M., so we need not show earlier afternoon hours than that. Combining these two restrictions warns us that the sun can strike our dial only between 4:30 and 5:00 A.M.; and again between 5:00 and 7:30 P.M. We might well decide to show on our finished dial the morning hour lines for 4 and 5 o'clock, and the evening hours from 5 to 8 inclusive. These hour lines have been drawn in solid in Figure 10.7, but the other hour lines, even if used in our intermediate computations, are shown as broken lines to signify that they will not be traced off onto our final dial plate.

After our dial has been completed and attached to the wall, with the gnomon in place, we will get an arrangement similar to that shown on the left side of Figure 10.8. This figure shows the cross section of a vertical wall viewed from the west looking east, with north at the left and south at the right. Vertical dials have been attached back to back to the opposite faces of the wall. The dial at the left is a north decliner and that at the right a south decliner. The styles of the two gnomons are parallel to each other and to the earth's axis. People sometimes loosely describe the situation by saying that "the gnomon of a northerly dial points up, while the gnomon of a southerly dial points down."

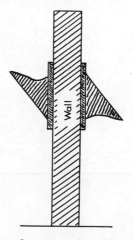

FIGURE 10.8 Gnomons of corresponding dials on opposite sides of a wall.

Measuring a Wall's Declination. Figure 10.9 shows two walls. In each case line *OB* lies in the meridian with *B* at the south end, and *OA* is perpendicular to the face of the wall. And in each case *BOC* is a right angle. For either wall the amount of the declination is equal to the angle *AOB* or the angle *DOC*. For either wall, we need to measure one of these two angles to know the amount of the declination. The upper wall declines toward the southwest and the lower one toward the southeast.

Perhaps the simplest way of measuring the angle of declination is to place a horizontal table or shelf against the wall, hold up a plumb line just when the sun souths (see Chapter 2), and mark the position of the noon shadow. Since at the moment of local apparent noon the sun is due south, the shadow of the plumb line will lie in

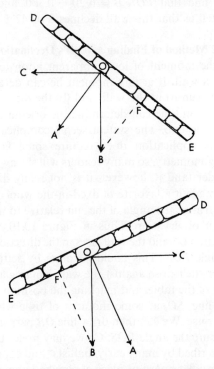

FIGURE 10.9 If arrow *OB* points toward the south and *AO* is perpendicular to the wall, the declination of each wall is measured by angle *AOB* or by angle *DOC*. If moreover *FB* is perpendicular to the wall, the declination is equal to angle *FOB*.

the meridian and if we mark its position on the horizontal table it will give us line *OB* of Figure 10.9. If we draw line *FB* perpendicular to the face of the wall, the declination can be measured from angle *FBO*. Let us measure on the tabletop the distance *FO* and the distance *FB* and divide the former by the latter. This will give us the tangent of the wall's declination. To be specific, in the upper part of Figure 10.9, we measure the distance *FO* on the table top and find that it is 8 inches and measure *FB* and find that it is 10 inches. We divide $FO/FB = 8/10 = 0.800$ which is the tangent of the wall's declination. Any trigonometric table will tell us that the angle whose tangent is 0.800 is 38°40′, so, since the wall itself faces toward the southwest, we say that it declines S 38°40′ W. If, in the lower part of Figure 10.9, our measurements show that *FO* is 6 inches and *FB* is 20 inches, we find that $FO/FB = 6/20 = 0.300$, and our trigonometric tables tell us that this wall declines S 16°42′ E.

An Alternative Method of Finding a Wall's Declination.   One need not wait for the moment of local apparent noon to ascertain the declination of a wall. If at any moment he can determine: (a) the direction of the sun from the wall, and (b) the direction of the sun from the south point, he can determine the orientation of the wall by comparing the two. The system seems complicated when described, and its application does require some familiarity with elementary trigonometry, so many readers will shy away from it. For those who understand it, however, it is not really difficult, and its accuracy has made it a favorite of dyed-in-the-wool diallists.

We first determine the angle of the sun relative to the wall, noting the time of our observation. Thus in Figure 10.10, where the sun lies in the direction *OS* and the wall faces in the direction *OB*, we want to find the angle *BOS*. This could be done by setting a carefully leveled table or shelf close against the wall, hanging a plumb line at the outer edge of the table, and marking the position of the shadow of the plumb line, *SO*, at some moment of time which we would record for later use. We can then draw line *OB* perpendicular to the wall, and measure the angle *BOS*. Or we may prefer to use a simple instrument described by many early diallists[4] and depicted in Figure 10.11. *ABCD* is a flat board or piece of plywood about a foot square with a nail, *EF*, driven into the board a little below the top and

[4] See, for example, N. Bion, *De la Construction & des usages des cadrans solaires* (Paris: 1752).

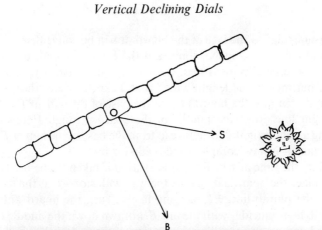

FIGURE 10.10 The wall faces in the direction *OB*, with the sun in the direction *OS*. Measuring angle *BOS* we find in this case that the sun lies 53½° to the left of a line perpendicular to the wall

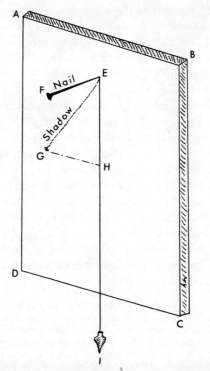

FIGURE 10.11 Simple instrument for measuring the angle of the sun.

perpendicular to the face of the board. It will be convenient to use a nail about $2\frac{1}{2}$ to 3 inches long, and the exposed length of the nail must be carefully measured. (Don't use the entire length of the nail, but merely the length which is still exposed after the nail has been driven into the board.) From the base of the nail, at *E*, draw a straight line down the middle of the board through *H* toward *I*. Hang a light weight, *I*, by a thread from the base of the nail at *E*. The instrument is now complete and ready for use.

Note the time at which the observation is taken (to be used later) and place the board flat against the wall with side *AB* at the top and with the plumb line, *EI*, hanging freely. Turn the board until the plumb line coincides with the line *EH* drawn down the middle of the board and mark carefully the position of the end of the shadow of the nail at *G*. Measure the distance, *GH*, from *G* to the plumb line, making the measurement perpendicular to the plumb line. We now have two measured distances, *EF* (the length of the nail), and *GH* (the displacement of the end of the shadow from the vertical). Divide *GH* by *EF*, and the quotient is the tangent of the angle of the sun from the wall. If, for example, the nail is 6.8 centimeters long, and if distance GH is 5.3 centimeters, then 5.3/6.8 = 0.779 is the tangent, and the corresponding angle turns out to be 37.9°, which would be the size of angle *BOS* in Figure 10.10. In this case, if we stand with our back to the wall looking out at right angles from the wall (as along line *OB* of Figure 10.10) the sun will be 37.9° to our left. We can judge this by the fact that in Figure 10.11 the tip of the nail's shadow is to the left of the vertical line as we face the wall. Figure 10.12 shows our instrument actually in use.

We now have the first of our requisites—the direction of the sun relative to the wall. To compare with it, we still need to know the sun's direction from the south point of the horizon. This direction is called the sun's *azimuth from the south*. If we are fortunate enough to have access to the proper published tables,[5] we can look up the sun's azimuth for any combination of latitude, solar declination, and local apparent time.

[5] *Tables of Computed Altitude and Azimuth*, U.S. Hydrographic Publication No. 214, available at about \$4–\$5 per volume from the Superintendent of Documents, U.S. Govt. Printing Office, Washington, D.C. 20402. Each volume covers 10° of latitude, so one should make sure to secure the volume for the desired latitude. The tables list azimuths from the *north*, but subtracting them from 180° yields azimuths from the south. Even rough interpolation in these tables will usually give results of usable accuracy.

FIGURE 10.12 Homemade instrument being used to determine the declination of a wall.

Lacking tables, we compute the sun's azimuth, starting with the clock time of our observation. We convert this to L.A.T. by methods covered in Chapter 2, and state the L.A.T. as an hour angle from noon. Thus if the L.A.T. is 9:36 A.M., this, being 2^h24^m before noon, gives an hour angle of 36°00.0'. (See Table A.5 of our Appendix for conversion from time to arc.) We shall call this value "t" in our computations. We must also know the value of the sun's declination at the time of our observation. This value, which we shall call "dec," can be taken from Table A.2 of our Appendix or, more accurately, from a current ephemeris[6] or from many current

---

[6] Simplest and least expensive is the annual paperback leaflet, *Ephemeris of the Sun, Polaris, and Other Selected Stars*, prepared by the Nautical Almanac Office of the U.S. Naval Observatory and available from the Superintendent of Documents, U.S. Government Printing Office, Washington, D.C. 20402 for approximately 35¢.

almanacs. And finally we must know the latitude, ϕ, of the place of our observation. Let us assume that, for our particular observation, we have these values:

$$t = 36°00.0' \qquad \text{dec} = 17°46.7' \qquad \phi = 41°48.5'$$

We proceed to find the sun's declination by using these three values in solving two equations:

(1) $\log \tan M = \log \tan \text{dec} - \log \cos t$
(2) $\log \tan Z' = \log \tan t + \log \cos M + \text{colog} \sin (\phi - M)$

In these equations, Z' is the sun's azimuth from the south, and M is merely an intermediate value for use in the second equation. We will find it useful to organize our computations in tabular form, as in Table 10.3. The value of Z' is the sun's azimuth, or its angular distance from the south point. Since our observation was taken before local apparent noon, the sun was *east* of the meridian by 62°56.1'.

TABLE 10.3
DETERMINING THE SUN'S AZIMUTH

dec	(17°46.7')	log tan dec	9.50603		
t	(36°00.0')	− log cos t	9.90796	log tan t	9.86126
M	(21°37.2')	log tan M	9.59807	log cos M	9.96832
				colog sin $(\phi - M)$	0.46205
ϕ	(41°48.5')			log tan Z'	0.29163
M	(21°37.2')			Z'	62°56.1'
$(\phi - M)$	(20°11.3')				
log sin $(\phi - M)$		9.53795			

We have now seen that the sun was 37.9° to our left as we faced out from the wall (we shall call this value "W"), and the sun's azimuth, Z', at the same moment was 62°56.1' east of south. We have the situation shown in Figure 10.13, where AB represents the wall, with the sun off to the left toward OC, with angle $WOC = 37.9°$. But at this moment the sun was 62.9° east of south, so we lay off angle $COS = 62.9°$. The angle WOS, which is the wall's declination, is the difference between them, or 25.0°, and we discover that our wall faces S 25.0° E.

We are well-advised in all such problems to draw a rough diagram similar to Figure 10.13; but experimentation with such diagrams

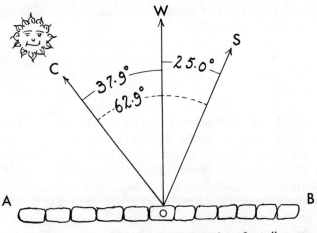

FIGURE 10.13 Computing the orientation of a wall.

will quickly show the validity of the following rules. Let W represent the angle between the sun and the wall (37.9° in our case and in Figure 10.13) and let Z' represent the sun's azimuth from the south (here 62.9°). Then:

(A) If the observation was before local apparent noon:

  (1) If the sun was to the right of the wall:
    (a) The wall declines *east* by the amount $Z' + W$.

  (2) If the sun was to the left of the wall:
    (a) If W is greater than Z', the wall declines *west* by their difference, $W - Z'$.

    (b) If Z' is greater than W, the wall declines *east* by their difference, $Z' - W$.

(B) If the observation was made after local apparent noon:

  (1) If the sun was to the left of the wall:
    (a) The wall declines *west* by the amount $Z' + W$.

  (2) If the sun was to the right of the wall:
    (a) If W is greater than Z', the wall declines *east* by their difference, $W - Z'$.

    (b) If Z' is greater than W, the wall declines *west* by their difference, $Z' - W$.

In our illustrative case, with a morning observation and the sun to the left (as we face straight out from the wall), and with $Z' = 62.9°$

and $W = 37.9°$, the wall declines from the south toward the east by the difference of 25.0°.

We must keep in mind that the sun's declination in winter is negative, so that in solving our formulas dec will be negative and M will be negative, and in getting the value of $\phi - M$ we subtract algebraically.

Dials with Large Declination. The typical vertical sundial has hour lines which radiate from the dial center—at the top of southerly dials and at the bottom of northerly dials. But direct east or west vertical dials have no centers. Their hour lines are parallel. And when the dial's declination is very large, getting close to 90°, the angles between the hour lines become very small, and the lines are difficult to distinguish if we lay out the dial in the ordinary fashion. We need some way to separate the hour lines.

Suppose, for example, that we are designing a dial for the vertical wall of a garage which faces S 82° W in Stonington, Connecticut (latitude 41°20′ N). Methods described earlier in this chapter tell us that:

Sub-style distance is	48°23.4′
Style height is	5°59.9′
Difference in longitude is	84°41.8′
Angle of 6 P.M. line is	48°56.7′

The difference in longitude tells us that the sub-style lies between the hour lines of 5 P.M. and 6 P.M.

Since this dial faces almost due west, there will be no use for most of the morning hour lines. Our methods described before tell us that the sun will strike this dial only between the hours of about 11:30 A.M. and 7:30 P.M.; so we shall work only with the hour lines from 12 noon to 8 P.M. inclusive. Still using methods described earlier in this chapter, we find the angles which the hour lines make with the sub-style, and since we know that the sub-style itself makes an angle of 48°23.4′ with the vertical, we can find the angle which each hour line makes with the vertical. The results appear in Table 10.4.[7] Inspection of the values shown in the table will show at once that the hour lines are crowded close together. The angles between them are very small. In fact, if we were to show the hour lines for 1 P.M.,

---

[7] The reader may carry out the calculations to check his understanding of the methods, seeing if his results correspond with these.

TABLE 10.4
ANGLES OF HOUR LINES ON A DIAL OF LARGE DECLINATION

hour line	angle from		hour line	angle from	
	sub-style	vertical		sub-style	vertical
NOON	48°23.4′	0°00.0′	5 P.M.	1°01.4′	47°22.0′
1 P.M.	15°46.4′	32°37.0′	sub-style	0°00.0′	48°23.4′
2 P.M.	8°23.7′	39°59.7′	6 P.M.	0°33.3′	48°56.7′
3 P.M.	4°57.5′	43°25.9′	7 P.M.	2°12.9′	50°36.3′
4 P.M.	2°00.1′	46°23.3′	8 P.M.	4°13.9′	52°37.3′

3 P.M. and 8 P.M. they would fall like the lines of Figure 10.14, and we will realize that most of the remaining hour lines would have to be crowded between them. The lines would lie so close together that it would be difficult to read the time from them. We must find some way to spread the hour lines farther apart. But it occurs to us that as we move farther away from the dial center, the lines do get farther and farther apart, and since there is no reason that we need to show the

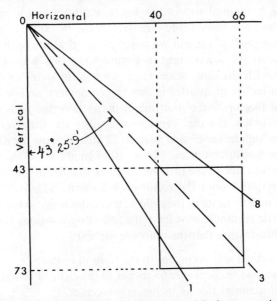

FIGURE 10.14 Hour lines for 1 o'clock and 8 o'clock on a vertical dial facing S 82° W in latitude 41° 20′ N. The 3 o'clock line is also shown as a broken line. The rectangle at the lower right shows the part of the field chosen for use in the completed dial.

dial center on our finished dial, it occurs to us that we might use merely that part of Figure 10.14 represented by the rectangle at the lower right. In that case, the hour lines will nearly fill the rectangle.

First we must decide on the size of our finished dial. This will naturally depend on its location. Let us suppose that the dial plate is to be rectangular measuring 36 inches horizontally and 40 inches vertically. (Any other dimensions would do as well, but we must start with *some* figures.) Let us further suppose that there is to be a margin 5 inches wide clear around the rectangle, within which we will ultimately place the numbers of the hour lines. Cutting off 5 inches from each side leaves us a smaller rectangle 26 inches wide and 30 inches high which will contain the hour lines. We draw a rectangle to scale, with its sides in the proportions of 26 to 30. These sides need not measure 26 and 30 inches, but may be 26 quarter inches and 30 quarter inches, or 26 centimeters and 30 centimeters, or in any other units which will give us a rectangle of convenient size to move about on Figure 10.14. We place our little cutout rectangle on Figure 10.14 and move it around, keeping the 26-inch side parallel with the horizontal top of the figure, until we find a position at which the hour lines approximately fill the rectangle. There is no single "correct" place, but we might decide that the position of the rectangle at the lower right in Figure 10.14 is a good one. If we measure with the same scale which we used in laying out the rectangle (in inches or quarter inches or centimeters, and so forth) we find that the top horizontal line of the little rectangle is now, say, 43 units below the dial center—and since the little rectangle is 30 inches high its lower side must be 73 inches below the dial center. Similarly measurement on Figure 10.14 might show that the left-hand vertical side of the little rectangle is 40 units to the right of the dial center, and since the rectangle is 26 units wide its right-hand edge must be 66 units to the right of the dial center. So the four sides of the little rectangle now have the following positions (we assume for our illustration that the units are inches):

(a) A vertical side 40 inches to the right of center.
(b) A second vertical side 66 inches to the right of center.
(c) A horizontal side 43 inches below center.
(d) A second horizontal side 73 inches below center.

Starting with these distances, and knowing the angles which the hour lines make with the vertical (and hence with the horizontal), it is but an exercise in elementary plane geometry to find where each hour line intersects the sides of the rectangle. We shall carry out the computation for one sample hour line in detail as an illustration and merely tabulate the results for the others. For our detailed sample we shall use the hour line of 3 P.M.

The 3-o'clock hour line, as we saw in Table 10.4, makes an angle of 43°25.9′ with the vertical. Trigonometric tables tell us that the tangent of 43°25.9′ is 0.9467, so for every inch which we drop vertically below the dial center this hour line moves 0.9467 inches toward the right. By the time we have dropped the 43 inches to the level of the top of our rectangle we shall have moved 43 times 0.9467 or 40.71 inches to the right; but since the left-hand edge of the rectangle is already 40 inches to the right of the dial center, we shall have passed it by 0.71 inches, and the hour line of 3 P.M. will cross the top edge of the little rectangle 0.71 inches from its left end.

If we carry out the same processes to find where this hour line intersects the bottom of the rectangle, we will find that by the time the hour line has dropped the 73 inches to the level of the lower edge of the rectangle it will have moved 69.10 inches to the right of the dial center, which is 3.10 inches beyond the far side of the rectangle. Hence this hour line must cross the right-hand vertical side of the rectangle before it "reaches bottom." But since the cotangent of the angle of the 3-o'clock line (that is, cot 43°25.9′) is 1.0563, we know that every time we move one inch to the right of the dial center the hour line drops 1.0563 inches; and by the time we have moved the 66 inches to the right to reach the right-hand side of the little rectangle we will have dropped 69.72 inches. The top of the rectangle is already 43 inches below the dial center, so we have dropped 26.7 inches below the top, and the 3 P.M. hour line will intersect the right-hand side of the rectangle at a point 26.7 inches down from the top. Having found where this hour line crosses the top and the right side, we merely connect these two points with a straight line and have the desired 3-o'clock hour line. We proceed similarly to find the points where the remaining hour lines cross the rectangle, giving the results in Table 10.5. In Table 10.5, "DL" means "down the left side of the rectangle measured from the top," "DR" means "down the right side from the top," "AT" is "across the top measured from the left side" and "AB" is "across the bottom from the left." We thus have two points for each hour line, and connecting these two points with a straight line gives the hour line itself. The sub-style, which

TABLE 10.5
LOCATING HOUR LINES ON A DIAL OF LARGE DECLINATION

hour line	left end	right end	hour line	left end	right end
1 P.M.	19.51 DL	6.72 AB	5 P.M.	6.71 AT	17.76 DR
2 P.M.	4.68 DL	21.24 AB	6 P.M.	9.37 AT	14.48 DR
3 P.M.	0.71 AT	26.72 DR	7 P.M.	12.36 AT	11.21 DR
4 P.M.	5.14 AT	19.88 DR	8 P.M.	16.29 AT	7.43 DR

makes an angle of 48°23.4′ with the vertical, is located in the same way, and crosses the rectangle at 8.42 AT and 15.62 DR. The hour lines and the sub-style are shown in Figure 10.15.

We must still design the style itself. We have found the two points where the sub-style intersects the rectangle. The upper intersection is 8.42 inches across the top of the rectangle, which places it 43 inches below and 48.42 inches to the right of the dial center. The straight-line distance from the dial center can be found by the Pythagorean rule of "the square on the hypotenuse":

$$48.42^2 = 2344.50$$
$$43^2 \quad = 1849.00$$
$$\text{sum} = 4193.50$$

The square root of 4193.50 is 64.76, which means that the point where the sub-style crosses the top of the rectangle is 64.76 inches from the dial center. But we found that the style height is 5°59.9′. The tangent of 5°59.9′ is 0.1051, so the style will rise 0.1051 inches for every inch away from the dial center, and when we get 64.76 inches from the dial center the height of the style will be 6.80 inches above the dial plate. Using similar reasoning we will find that when the style reaches the point where it crosses the right-hand side of the rectangle, its height above the dial plate will be 9.28 inches, and that the entire length of the sub-style line within the boundaries of the rectangle is 23.51 inches. So we will have a gnomon shaped like that at the bottom of Figure 10.15, 23.51 inches long across the bottom, 6.80 inches high where it crosses the upper edge of the rectangle, and 9.28 inches high where it crosses the rectangle's right-hand side. We set this gnomon vertically on the sub-style line and attach the sundial to the wall with the upper edge horizontal, and we are "in business." Figure 19.2 on page 193 shows an actual dial facing S 80°51′ E which was constructed by these methods.

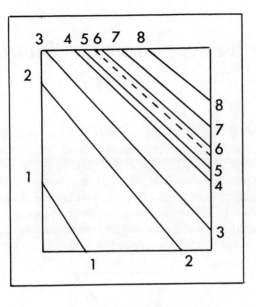

FIGURE 10.15 Vertical sundial for wall declining S 82° W in latitude 41° 20′ N. The sub-style is shown as a broken line. The gnomon, drawn to the same scale, appears at the bottom. The dial center is far off the diagram at the upper left.

11
Direct Reclining or Inclining Dials

The preceding chapters have considered horizontal dials and all types of vertical dials. We have also discussed two types of reclining dials: the equatorial and the polar dial. Reclining and inclining dials are placed on sloping surfaces, such as the roof of a house. In Figure 11.1, a horizontal dial would be placed on surface *A* and a vertical dial on surface *D*. The sloping surface, *C*, which faces upward toward the sky, would take a reclining dial and the sloping surface, *B*, which faces downward toward the ground, would take an inclining dial.

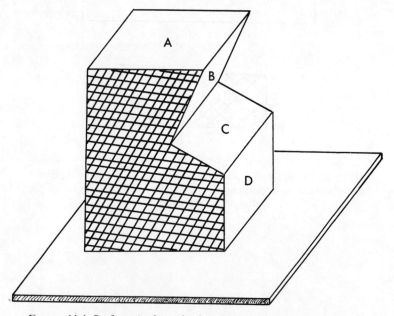

FIGURE 11.1 Surface *A* takes a horizontal dial, surface *B* an inclining dial, surface *C* a reclining dial and surface *D* a vertical dial.

Neither type of dial is common, although for obvious reasons recliners are more common than incliners.

Reclining and inclining dials are called *direct* if their planes face any one of the four cardinal points of the compass. We discover whether or not a given surface is direct by drawing a horizontal line on the surface (using a carpenter's level), and then drawing a second line on the surface perpendicular to the horizontal line. If this second line lies in the meridian, the plane is a direct plane and will carry a direct north or south reclining or inclining dial. If the second line lies in the prime vertical (that is, if it runs due east and west) the sloping surface will carry either a direct east or a direct west reclining or inclining dial. In this chapter we shall treat only those sloping dials which are direct. Polar dials, described in Chapter 9, are direct south recliners.

A plane's *reclination* is the angle which the plane makes with the vertical, and can be measured with a plumb line or a carpenter's level. If we have marked on our sloping surface the horizontal line mentioned in the preceding paragraph, and also the second line perpendicular to it, we now place the edge of a board on the roof along this second line, with the board itself perpendicular to the roof, as shown in Figure 11.2. If the line *AB* is parallel to the roof, and if we hang a plumb line from *C* in the upper part of the figure, the angle which the plumb line makes with *AB* (that is, angle *AOC*) is the reclination of the plane. Or if we use a carpenter's level, as in the lower part of the figure, with *AB* again parallel to the slope and *EF* the carpenter's level, then angle *FOB* is the complement of the plane's reclination.

Direct South or North Recliners or Incliners. If the plane on which the dial is to be placed faces exactly north or south our problem is merely that of "reducing the plane to a new latitude," and then designing a horizontal dial for that new latitude. We shall adapt rules for "reducing to a new latitude" from those given by Captain Samuel Sturmy nearly 300 years ago.[1] They run as follows:

(a) Find the latitude of the place where the dial is to be used.

(b) Compute the colatitude by subtracting from 90°.

(c) Find the amount of reclination or inclination of the plane on which the dial is to be placed.

[1] Capt. Samuel Sturmy, *The Art of Dialling* (London: 1683).

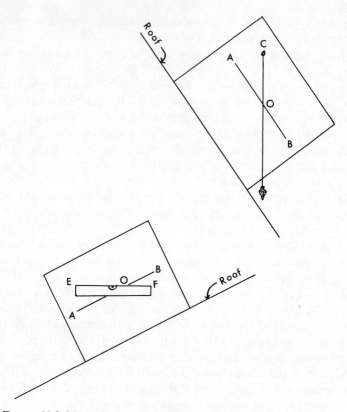

FIGURE 11.2 Measuring the reclination of a roof by means of a plumb
line (upper) or a carpenter's level (lower).

(d) Then proceed as follows:

(I) For south recliners or north incliners:

(a) If the reclination (or inclination) and the colatitude are
unequal, subtract the smaller from the larger, and the difference is
the new latitude for which you design a horizontal sundial.

(b) If the reclination (or inclination) and the colatitude are
equal, design a polar dial by the methods of Chapter 9.

(II) For south incliners or north recliners:

(a) The sum of the reclination (or inclination) and the colatitude
is the new latitude—or, if this sum exceeds 90°, subtract it from
180° to find the new latitude. Design a horizontal dial for the new
latitude.

(b) If the reclination (or inclination) and the latitude are equal, design an equatorial dial. (See Chapter 4.)

If the reader has any doubt about measuring the inclination of a plane, he can remember that the inclination of the under side of a plane is exactly equal to the reclination of the upper side.

To illustrate our rules:

Example 1 A direct south plane reclining 20° in latitude 43°. Colatitude = 47°. Use Rule Ia. Design a horizontal dial for latitude 27°.

Example 2 A direct south plane reclining 51° in latitude 39°. Colatitude = 51°. Under Rule Ib, design a polar dial.

Example 3 A direct north dial reclining 10° in latitude 36°. Colatitude = 54°. Under Rule IIa design a horizontal dial for latitude 64°.

Direct East or West Recliners and Incliners. These planes are reduced to new latitudes for which we design vertical south declining dials by the methods of Chapter 10. Let us speak of the latitude in which the dial will actually be used as the *dial latitude*, and the new latitude for which we will design a vertical south decliner as the *reduced latitude*. Then we can state the following rules:

(I) The reduced latitude is the complement of the dial latitude, found by subtracting the dial latitude from 90°.

(II) The declination of the equivalent vertical south declining dial is the complement of the reclination (or inclination) of the sloping plane on which the dial is to be placed.

(III) On the final dial, the 12-o'clock line will be a horizontal line with the dial center at one of its ends, in accordance with the following rules:

(A) With direct east recliners, the 12-o'clock line is at the bottom and the dial center at the left.

(B) With direct east incliners, the 12-o'clock line is at the top and the dial center at the right.

(C) With direct west recliners, the 12-o'clock line is at the bottom and the dial center at the right.

(D) With direct west incliners, the 12-o'clock line is at the top and the dial center at the left.

(IV) Having found the equivalent vertical south declining dial using the reduced latitude from Rule I and the declination of Rule II, compute the values of SD, SH, DL, and AV and the positions of the hour lines by the methods of the preceding chapter, and lay them out measured from the horizontal 12-o'clock line and dial center specified under Rule III. This will give the required dial.

We illustrate the application of these rules by designing a sundial to be placed on a direct east surface reclining 66° in latitude 50°.

(1) The reduced latitude is $90° - 50° = 40°$.

(2) The declination is $90° - 66° = 24°$.

(3) The 12-o'clock line is at the bottom with the dial center at the left.

(4) We compute the basic values for a vertical dial declining S 24° W in latitude 40°. We have already computed the basic values for this dial in Table 10.1, where we found that SD = 25°51.7′; SH = 44°24.7′; DL = 34°42.5′; and AV = 71°09.4′. The style will lie between the hour lines of 9 A.M. and 10 A.M.

(5) Lay out the 12-o'clock line (see Figure 11.3) near the bottom with the dial center at the left, and lay off the sub-style radiating from the dial center making an angle with the 12-o'clock line equal to SD (here 25°51.7′).

(6) The hour lines (whose angles are already tabulated for this case in Table 10.2 on page 80) are now laid out at the proper angles

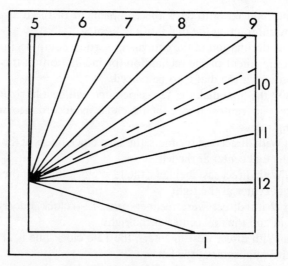

FIGURE 11.3 Direct east dial reclining 66° from the vertical in latitude 50°.

from the sub-style, keeping in mind that the hour lines for 10 A.M. and 11 A.M. will lie between the sub-style line and the 12-o'clock line, while the hour lines for 9 A.M., 8 A.M., and so forth will lie on the other side of the sub-style.

It will help us to understand the layout of hour lines on these dials if we keep in mind the fact that any direct west dial will show mainly afternoon hours, and any direct east dial will show mainly morning hours. Our 12-o'clock line is always horizontal, and the dial center always at one end.

12
Dials Which Both Decline
and Recline

It is unlikely that the reader will have occasion to design a sundial for a surface which both declines and reclines. There are many such surfaces, of course, but they are seldom used as supports for sundials, partly because they are seldom convenient for the purpose and partly because of the complications in designing them. Early editions of the *Encyclopedia Britannica* gave full trigonometric directions for such dials, but nearly a century ago this material was eliminated and detailed directions retained for only the simpler dials. Commenting on the increased likelihood of error in designing declining reclining dials, the *Encyclopedia* stated, "In all these cases . . . the only safe practical way is to mark rapidly on the dial a few points (one is sufficient when the dial face is a plane) of the shadow at the moment when a good watch shows that the hour has arrived, and afterward connect these points with the center by a continuous line. Of course, the style must have been accurately fixed in its true position before we begin." These cautions are as valid today as they were a century ago, and one cannot honestly advise the reader to go to the trouble of designing a dial for a declining reclining site unless he does it for pure enjoyment of the mathematical exercise. For those who do want such a dial we recommend the *Encyclopedia*'s method—the empirical one of setting up the style and drawing the hour lines by observation of the shadow rather than by calculation.

Of course, this practical method does require one to set up the gnomon correctly so that the style will be parallel to the earth's axis. Again, this could be done by complicated calculation, but if we are to approach the problem empirically we may as well approach the placement of the gnomon the same way. We illustrate with Figure 12.1. In this figure *ABCD* represents a sloping plane surface, like the roof of a house. Draw a horizontal line *EOH* on the roof, using a carpenter's level to locate it. Then *EFGH* is a flat board with

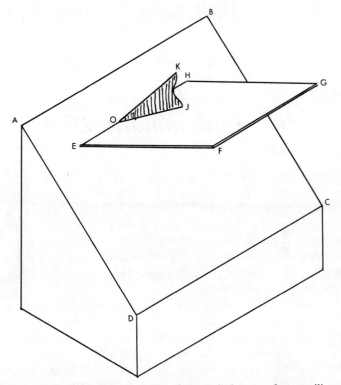

FIGURE 12.1 Finding the position of the style by use of an auxiliary level board.

its edge *EOH* pressed against the horizontal line on the roof and with the entire board carefully leveled. On the level board draw a section of the meridian, *OJ*, with *O* toward the north and *J* toward the south. On *OJ* erect perpendicularly the gnomon, *KOJ*, with angle *KOJ* equal to the latitude. Thus the horizontal board and the gnomon will be the dial plate and gnomon of a horizontal dial for the latitude. The style, *OK*, is now in the right position for any plane which passes through point *O*—not only for the horizontal plane, *EFGH*, but also for the sloping declining plane *ABCD*. If we retain the line *OK* in position and remove the board *EFGH*, we can continue the plane of the gnomon downward until it meets the roof, *ABCD*, and, attaching the gnomon to the sloping roof, we are ready to draw in the hour lines by noting the position of the shadow of *OK* on the roof at the desired hours.

13
The Analemmatic Dial

The style of most sundials lies parallel to the earth's axis, but there are a few exceptions to the rule. For example, the gnomon of a shepherd's dial (see Chapter 16) is a horizontal pin; and the gnomon of the analemmatic dial, which we now describe, is a vertical pin or rod which is moved about from place to place according to the sun's declination. Moreover, the analemmatic dial does not have hour *lines* in the usual sense, but hour *points* which fall along the circumference of an ellipse.

Analemmatic dials are comparatively rare, and there seems to be a feeling that they are difficult to make.[1] Actually their construction is no more difficult than that of an ordinary horizontal dial, and the two may be combined on the same dial plate for use together. Construction by graphic means, while quite possible, is, to be sure, rather complicated, so we will omit the graphic approach here and confine our treatment to a computational one.

Laying Out the Hour Points. Following Figure 13.1:

(1) Lay off AB and CD mutually perpendicular and intersecting at O. The point O does not show in the diagram because it is obscured by the central scale of dates; but we shall continue to refer to point O as a point from which to measure.

(2) Take AB as the major axis of the ellipse along which the hour points will fall, and designate by M the length of the semi-major axis, AO or OB. Let $M = 1.000$.

(3) The semi-minor axis of the ellipse, m, will be OC with its length found from the relationship

$$m = \sin \phi.$$

---

[1] Mayall & Mayall, in their splendid *Sundials*, say, "More often than not when one finds how much work is entailed in making it, another type more easily constructed is substituted.

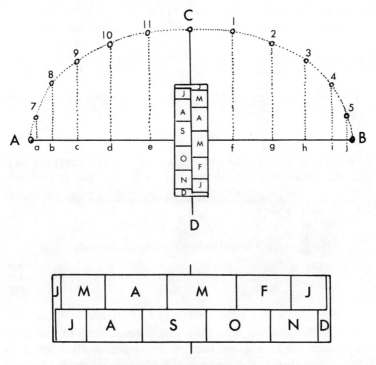

FIGURE 13.1 Analemmatic dial for latitude 41° 43′, with enlarged view
of scales of dates at the bottom.

If we lay out an analemmatic dial for Farmington, Connecticut, in
latitude 41°43′, we have

$$m = \sin 41°43′ = 0.665.$$

(4) Let the times of the various hour points be represented by their
hour angles from noon. That is, for example, the hours of 11 A.M. and
1 P.M. are each 1 hour from noon and have hour angles of 15°;
10 A.M. and 2 P.M. are 2 hours from noon and have hour angles of
30°, and so forth. Call this hour angle for any hour "t", and find the
horizontal distance, H, of the hour point from O by means of the
formula:

$$H = \sin t.$$

This gives us the horizontal distances of the various hour points for
every analemmatic dial in Table 13.1.

TABLE 13.1

ABSCISSAE (HORIZONTAL VALUES) FOR AN ANALEMMATIC DIAL

hour:	NOON	11 A.M.	10 A.M.	9 A.M.	8 A.M.	7 A.M.	6 A.M.
		1 P.M.	2 P.M.	3 P.M.	4 P.M.	5 P.M.	6 P.M.
H:	0.000	0.259	0.500	0.707	0.866	0.966	1.000

(5) Find the vertical height, V, of each hour point above the major axis, AB by the relationship:

$$V = \sin \phi \cos t.$$

Although the values of H will be the same for every analemmatic dial, the values of V will obviously vary with ϕ, and in our case, with $\phi = 41°43'$, repeated solution of the formula yields the values for V in Table 13.2.

TABLE 13.2

ORDINATES (VERTICAL VALUES) FOR AN ANALEMMATIC DIAL

hour:	NOON	11 A.M.	10 A.M.	9 A.M.	8 A.M.	7 A.M.	6 A.M.
		1 P.M.	2 P.M.	3 P.M.	4 P.M.	5 P.M.	6 P.M.
V:	0.665	0.643	0.576	0.471	0.333	0.172	0.000

(6) We now use the values of H and V, computed in Tables 13.1 and 13.2, to locate the hour points themselves. Thus to locate the hour point for 10 A.M. we measure out horizontally 0.500 units from O, and then vertically upward 0.576 units to reach the hour point. Similarly to find the hour point for 5 P.M. we start at point O, measure horizontally 0.966 units and upward 0.172 units, and we have located our point. All other hour points are similarly located. Morning points lie at the left of OC and afternoon points to the right.

(7) If we want hour points before 6 A.M. or after 6 P.M. we merely keep in mind that the ellipse is symmetrical. The points for 5 A.M. and 4 A.M. respectively have values of H exactly equal to the points for 7 A.M. and 8 A.M. respectively, but lie *below* line AB by the V distances for 7 A.M. or 8 A.M. At the other end of the ellipse, the hour points for 7 P.M. and 8 P.M. are just as far below AB as the points for 5 P.M. and 4 P.M. are above it.

(8) Analemmatic dials vary tremendously in final size, from tiny portable pocket dials with $AB = 2$ inches up to huge garden dials where $AB = 16$–20 feet. We have said, for purposes of computation, that we will "let $M = 1.000$." If we want our finished dial to have a semi-major axis of 2 inches, we multiply every value which we have computed by 2 to get our results in inches; but if we want a finished

dial with a semi-major axis of 8 feet, we multiply every computed value by 8 to get results in feet. Specifically, we have found that the hour point for 9 A.M. is 0.707 units horizontally and 0.471 units vertically from O; but if we want our finished dial to have a semi-major axis of 6 feet we multiply both figures by 6 and find that the 9 A.M. point lies 4.242 feet horizontally and 2.826 feet vertically from the dial center at O.

We now have all the hour points located, but we still need to lay out the scale of dates in the center. The vertical gnomon is placed on this scale at various points depending on the day of the year, and its shadow then points at the hour point corresponding to the time of day. This little scale of dates, or zodiac, can be computed from the formula:

$$Z = \tan \text{dec} \cos \phi$$

where dec is the sun's declination on the given day. Table 13.3 gives the sun's declination on the first day of each month, and its extreme values at the times of the solstices; and beside these declination figures are the corresponding values of Z computed by our formula for $\phi = 41°43'$.

TABLE 13.3
COMPUTATION OF THE SCALE OF DATES FOR AN ANALEMMATIC DIAL

date	dec	Z	date	dec	Z
Jan. 1	$-23°08'$	-0.328	July 1	$+23°00'$	$+0.326$
Feb. 1	$-17°18'$	-0.239	Aug. 1	$+18°00'$	$+0.250$
Mar. 1	$-8°00'$	-0.108	Sep. 1	$+8°30'$	$+0.115$
Apr. 1	$+4°15'$	$+0.057$	Oct. 1	$-2°54'$	-0.039
May 1	$+15°00'$	$+0.206$	Nov. 1	$-14°00'$	-0.192
June 1	$+22°00'$	$+0.311$	Dec. 1	$-21°40'$	-0.305
June 21	$+23°26\frac{1}{2}'$	$+0.333$	Dec. 21	$-23°26\frac{1}{2}'$	-0.333

We now lay out the central scale of dates as follows:

(9) Starting at O in Figure 13.1, measure upward toward C for each of the distances in Table 13.1 which is labeled $+$, and measure downward toward D for each of the distances which is labeled $-$. Thus the line for May 1 will be 0.206 units upward from O, and the line for November 1 will be 0.192 units downward from O. These distances must be multiplied by the actual length of the semi-major

axis, M, to convert them to the units of our real problem, just as we converted values of H and V in step 8.

(10) Note that the months of June and December "run around the ends" of the scale. The extreme ends of the scales mark the positions of the vertical gnomon on the days of the solstices.

Hour points may be found for the half and quarter hours by taking appropriate intermediate values of t. These dials may also be constructed with built-in corrections for longitude by using values of t which include the longitude correction for our own location, as explained for horizontal dials in Chapter 5.

Finding the Hour Points from Central Angles. The positions of the hour points may be found even more quickly and easily by computing angles at the center of the dial. We first draw an ellipse of the desired dimensions, using perhaps one of the methods described in Chapter 19, and we then find the positions of the hour points on this ellipse by computing the central angles. The steps in the procedure, illustrated in Figure 13.2, are as follows:

(1) Lay out the ellipse with the semi-minor axis, m, and the semi-major axis, M, related as follows:

$$m = M(\sin \phi),$$

where ϕ is the latitude. We illustrate with computations for a dial at Santa Fe, New Mexico, where the latitude is $35°41'$, and hence $m = 0.583\,M$. In Figure 13.2, the semi-minor axis, CO, is 0.583 times the semi-major axis, OB.

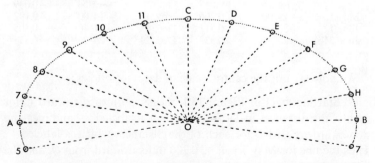

FIGURE 13.2 Laying out an analemmatic dial by means of the central angles.

(2) Compute the central angles for the various hour points using the following formula, in which t is the hour angle from noon:

$$\tan A = (\tan t)/(\sin \phi).$$

This formula is usually given in its logarithmic form:

$$\log \tan A = \log \tan t - \log \sin \phi.$$

The computations for this illustrative case are assembled in Table 13.4.

TABLE 13.4
COMPUTATION OF CENTRAL ANGLES FOR AN ANALEMMATIC DIAL IN
LATITUDE 35° 41'

time		t	$\log \tan t$	$\log \sin \phi$	$\log \tan A$	A
A.M.	P.M.					
11 or	1	15°	9.4281	9.7659	9.6622	24.7°
10 or	2	30°	9.7614	9.7659	9.9955	44.7°
9 or	3	45°	0.0000	9.7659	0.2341	59.7°
8 or	4	60°	0.2386	9.7659	0.4727	71.4°
7 or	5	75°	0.5719	9.7659	0.8060	81.1°
6 or	6	90°	∞	9.7659	∞	90.0°

(3) With the intersection of the major and minor axes at O as the center, and starting with the semi-minor axis, CO, lay off the angles of the last column of Table 13.4 and carry out the terminal sides of these angles until they reach the ellipse. Their intersections with the ellipse will mark the hour points. Thus, for example, angle COD is 24.7° and D will mark the hour point for 1 P.M.; angle COE is 44.7° and E will mark the hour point for 2 P.M. These same angles laid out on the opposite side of CO will locate the hour points of the morning hours.

(4) We now lay out the central scale of dates as explained in the preceding section. (See page 111.)

(5) The finished dial will not show the straight lines radiating from the dial center at O. These construction lines appear in Figure 13.2 merely to show which angles are laid out in finding the hour points. There are no hour *lines* in an analemmatic dial, which, in finished form, will look something like the dial shown in Figure 13.3.

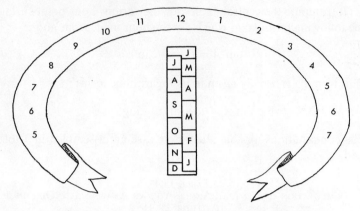

FIGURE 13.3 Analemmatic sundial for latitude 35° 41′ N.

General Observations. Since the analemmatic dial is a horizontal sundial, it should show hour points from the time of earliest sunrise to the time of latest sunset as shown in Appendix Table A.7. Analemmatic dials are sometimes combined with ordinary horizontal dials after the fashion of Figure 13.4. If such a combination dial is properly leveled and turned until the same time is shown on both dials, the dials will be automatically properly oriented.

The French astronomer, Lalande, suggested that one might lay out the ellipse of an analemmatic dial on a level garden spot, properly proportioned to the latitude, marking the hour points with beds of flowers and including the central scale of dates—but omitting the vertical gnomon. A visitor to the garden standing at the proper point on the scale of dates with his back to the sun would see his own shadow stretching across the lawn to show the time of day. Thus the sundial enthusiast becomes the gnomon for his own sundial! Lalande designed the famous analemmatic dial with a major axis measuring some 40 feet at the Church of Brou in the outskirts of Bourg-en-Bresse some 230 miles southeast of Paris, and another fine example is to be seen in Longwood Gardens at Kennett Square, Pennsylvania, 10 or 12 miles northwest of Wilmington, Delaware. Certainly the analemmatic dial deserves to be much more popular.

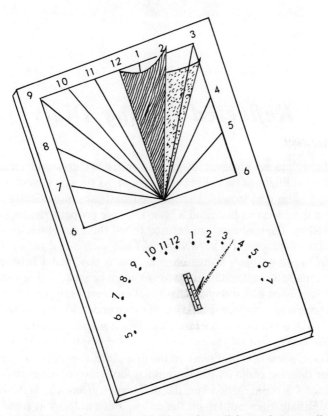

FIGURE 13.4 Analemmatic and horizontal sundials on the same dial plate, oriented so that both shadows indicate the same time.

14
Reflected Ceiling Dials

A mirror placed horizontally in a southern window will reflect a beam of sunlight to the ceiling of the room. As the sun moves across the sky from east to west, the reflected sunbeam moves across the ceiling from west to east; and if hour lines are properly drawn upon the ceiling, the sunbeam can be used to tell the time of day or even the day of the year. Isaac Newton laid out such a dial on a ceiling in his grandmother's home when he was a boy, and Christopher Wren made a similar dial in his home when he was a lad of sixteen. Any interested boy today can easily follow their example.

As one might guess, a dial reflected to the underside of a horizontal surface by a horizontal mirror is merely a horizontal sundial turned upside down, and all the rules of Chapter 5 apply to it. There are, however, a few new problems. In the first place, these dials are much larger than the ordinary garden sundial, since they cover most of the ceiling of a room. Moreover, the dial center, from which the hour lines radiate, does not fall on the ceiling, but outdoors at some distance from the house. Consequently our approach will have to differ in some ways from that which we have used for other horizontal dials.

We start by placing the mirror in a window of a room with southerly exposure and with a goodly expanse of ceiling on which the dial may be drawn. The mirror may be placed on the window sill, or perhaps on the upper ledge of the lower sash if it is a double-hung window. The higher it can be placed above the floor and the closer to the ceiling, within reason, the better, since the lower we drop the mirror the more we magnify the movement of the sunbeam on the ceiling, and consequently the fewer hour lines we can include. The mirror must be as nearly horizontal as possible, since the accuracy of the dial will depend on it.

Having the mirror in place, we next draw a meridian line on the ceiling, passing through the point vertically over the mirror. Thus if the upper part of Figure 14.1 represents a vertical cross section of the room, with the mirror on the window sill at *M*, our meridian line on the ceiling must pass through the point *S* which is vertically above the mirror. The meridian may be located by any of the methods des-

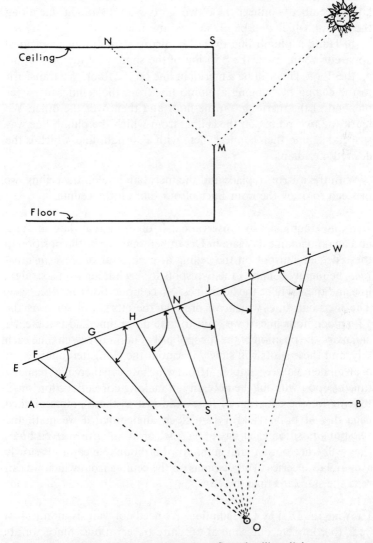

Figure 14.1 Laying out a reflected ceiling dial.

cribed in Chapter 3, but two methods are particularly well suited for reflected ceiling dials:

(a) Mark the positions of the reflected sunspot on the ceiling at the exact moment of local apparent noon on each of two days of widely different solar declination—say a day in late June and another in late December. Connect these two spots with a straight line along the ceiling, which will be the desired meridian.

(b) Hang a plumb line over the mirror at the moment of local apparent noon, mark the position of the shadow of the plumb line on the floor. This will be a meridian line on the floor. We transfer it to the ceiling by hanging a plumb line from the ceiling, first over one end of the meridian on the floor and then over the other. We mark the two points on the ceiling from which the plumb line was suspended, and then connect them with a straight line, which is the desired meridian.

With the mirror in place and the meridian line on the ceiling, we proceed to draw the hour lines of our dial on the ceiling.

Drawing Hour Lines by Observation. If on some day late in June, and again on some day late in December, one marks the position of the reflected sunspot on the ceiling at each hour of *local apparent time*, he merely connects the two spots for each hour with a straight line and these will be the hour lines. No computations are necessary. The meridian line, which we drew at the start, will serve as the 12-o'clock hour line. If we want to take more time and trouble, we can mark the position of the sunspot just at noon by watch time each day, and these spots will slowly trace out the elongated figure eight of an analemma (see Chapter 3) from which we can thereafter read the time of noon without the necessity of making corrections for longitude or for the equation of time. And just as these points marked each day at noon yield a 12-o'clock analemma, if we mark the sunspot's position each day at 11 A.M., or at any other given hour, they will trace out an analemma for that hour. We can thus slowly accumulate a series of analemmas on the ceiling, from which we can read the standard time directly.

Drawing the Dial by Computation. We need not wait six months or a year to complete our sundial by daily observations. Since we are designing an inverted horizontal dial we can design it by computation like any other horizontal dial. We already have the meridian on the ceiling, which will be the 12-o'clock line of the finished dial. We now find the remaining hour lines by the following steps:

(1) Measure carefully the vertical distance from the mirror at M to the point on the ceiling vertically above it at S (see the upper part of Figure 14.1). Call this vertical distance V.

(2) Starting at S in the figure, measure north along the meridian on the ceiling a distance

$$SN = (V)(\tan \phi),$$

where ϕ is the latitude. Thus if we are designing a dial for a house in Farmington, Connecticut ($\phi = 41°42.9'$), and if the vertical distance V is 60 inches, this becomes:

$$SN = (60)(\tan 41°42.9') = 53.49 \text{ inches.}$$

We locate point N on the meridian 53.49 inches from S. We could have found a close approximation to our value of SN by reference to Table A.11 in our Appendix, where the figures in the column headed "Equinox" are the appropriate values of $\tan \phi$.

(3) We now transfer our attention to the lower half of Figure 14.1, where we show the ceiling as it would appear if we were lying on the floor looking straight up at it. AB is the south wall of the room, and our meridian is SN. If the house faced exactly south the meridian would lie at right angles to the south wall. Here the house declines toward the east, and angle NSA is the complement of the declination. In the diagram, N is at the north, S at the south, E at the east, and W at the west, as will be apparent if we remember that we are looking upward at the ceiling. Points N and S of the lower diagram are the same as points N and S of the upper diagram, and the distance SN is 53.49 inches, as above.

(4) Through point N on the ceiling draw EW perpendicular to NS. This line EW is the equinoctial, along which the reflected sunbeam will move on the days of the equinoxes.

(5) We now want the points on the equinoctial, F, G, H, and so forth, where the various hour lines intersect the equinoctial. The morning sun will cast its reflected beam toward the west at L, and hour by hour it will move eastward through K, J, N, and so forth. We can think of the hour line through J as being the hour line of 11 A.M., that through H as being the hour line of 1 P.M., and so forth.

(6) Denote each hour by the sun's hour angle from local apparent noon. Thus the hour lines for 11 A.M. and 1 P.M. will both be 15° from noon; those for 10 A.M. and 2 P.M. will be 30° from noon, and so forth. Call this sun's hour angle for any desired hour t.

(7) The distance, d, along the equinoctial line from N to the intersection with any desired hour line, t, will be:

$$d = (V)(\tan \phi + \cot \phi)(\sin \phi)(\tan t).$$

For our dial in Farmington the value for the 10 A.M. or the 2 P.M. hour line would be:

$$d = (60)(\tan 41°42.9' + \cot 41°42.9')(\sin 41°42.9')(\tan 30°)$$
$$= 46.41 \text{ inches.}$$

It is not necessary, however, to carry out the calculation, since the results are given in Table A.12 in our Appendix. In that table, for example, we find that for $\phi = 42°$ the 10-o'clock line is 0.7771 (V) distant from the meridian; and since V is 60 inches this would give us (60)(0.7771) or 46.63 inches as compared with the 46.41 inches computed above—and had we interpolated between the tabulated values for latitudes of 41° and 42° we would have arrived at a value of 46.42 inches.

(8) The dial center from which the hour lines radiate will not lie on the ceiling, but rather at a point outdoors at the level of the ceiling on the extension of the meridian, like point O in Figure 14.1. But we learned in Chapter 5 how to find the angles which the various hour lines make at the dial center with the 12-o'clock line. There we learned that angle NOK for the 10-o'clock line at latitude 41°42.9' would be:

$$\tan NOK = \tan t \sin \phi = (\tan 30°)(\sin 41°42.9')$$
$$= 0.38418$$
$$NOK = 21°01.0'.$$

But if, in the right triangle KNO, angle NOK is 21°01.0', angle NKO must be the complementary angle $90 - 21°01.0'$ or 68°59.0'. At point K on the equinoctial we lay off angle NKO equal to 68°59.0' and draw our hour line for 10 A.M., extending it both ways from K even though only the southern part of it is shown in the figure.

(9) We continue with the other hour lines on the same basis, computing first the angle which the hour line makes with the meridian at the outdoor dial center at O, subtracting from 90° to get the complementary angle which this hour line makes with the

equinoctial, and then laying off the hour line at the proper angle through the point of intersection already found in step (7).

(10) Table A.11 of our Appendix gives values for finding where the dial center will be. Thus it tells us that in latitude 42° the dial center will lie on the meridian south of point N in Figure 14.1 by a distance of 2.011 V, or, in our example, where V is 60 inches, by (60)(2.011) or 120.66 inches. Since we have already discovered that NS is 53.49 inches, it is clear that the dial center lies outdoors at a distance of about 120.66 − 53.49 or 67.17 inches. Had we wished to find the more precise distance we could have used the formula that distance NO of Figure 14.1 is equal to $(V)(\tan \phi + \cot \phi)$, which would have given a distance of 2.0132 V or 120.79 inches. We do not need to find this dial center in laying out our reflected ceiling dial, although a knowledge of it was necessary in computing the tables of the Appendix.

Now that we have the hour lines and the equinoctial for our ceiling dial, we may add lines of solar declination such as the Tropics of Cancer and Capricorn by the methods described in Chapter 15. The diagram shown in Figure 14.1 is but a partial one, showing how to find those parts of the hour lines which lie south of the equinoctial; but all during the six months of the winter season when the sun is low the reflected sunbeam will fall farther back on the ceiling than line EW, and we must extend the hour lines farther toward the north to accommodate it. The little arcs with their arrows shown in Figure 14.1 will not be shown in the finished diagram. They are merely the angles which are to be measured in laying out the hour lines. And we should finally caution again against placing the mirror at too low a level. For example, if we place it on a window sill 60 inches below the ceiling (which is not an uncommon height for window sills) we will find that at the times of the equinoxes the hour lines for 9 A.M. and 3 P.M. will lie $6\frac{1}{2}$ feet either side of the meridian in latitude 40°, requiring a room 13 feet wide at the equinoxes and much wider in winter just for the six hours at midday. This expanse will be reduced proportionately as we reduce the distance from the mirror to the ceiling.

15
Dial Furniture

The lines and figures on a sundial which are used in telling the time of day are considered the basic or fundamental parts of its design, and all other lines or figures which are added to it, giving other sorts of information, are called *dial furniture.* Thus Figure 15.1 shows the face of a dial designed some three centuries ago for the latitude of London—a dial which has become so complicated that none but an expert can read it. In fact, the author of one of the chapters of the book in which this diagram first appeared prefaced his directions for designing dial furniture with the admonition,[1]

> Though to speak my own judgement, I think these kind of Additions rather for Ornament than use.... [This is in part] because the multitude of lines often hinders those that are not used to them, to tell the Hours of the day, which is the chief use of Sun Dials.

Similarly Diderot's Encyclopedia,[2] after describing the methods of finding the usual hour lines, warns us:

> Further description of various kinds of lines or points is based more on curiosity than on usefulness; the essential requisites of a good sundial are that the hour lines, and especially the meridian line, be accurately drawn, and that the style be properly set, and all the others lines which one might describe for showing things other than the time of day can at times bring disadvantage from confusion.

If one is designing a sundial for use by others he will usually be well advised to keep it simple, but we would encourage anyone who is making a dial for his own use to include such furniture as pleases him, since he will presumably understand that which he has himself

---

[1] William Leybourn, *Dialling* (London: 1682). The quotation is from the "Tenth Tractate" of the book, written by John Twysden.

[2] Denis Diderot, *Encyclopedie ou dictionnaire des sciences* (Geneva: 1777).

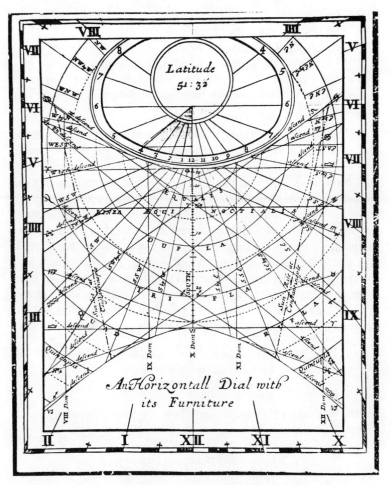

FIGURE 15.1 Furniture of an old horizontal dial.

designed. Types of sundial furniture which sometimes appear are (and this list is not complete):

(1) The maker's name and date.
(2) A motto.
(3) The equation of time.
(4) The time in other cities or countries.
(5) The sun's azimuth or direction.
(6) The sun's altitude.

(7) The day of the year (a calendar).

(8) The times of sunrise and sunset.

(9) The length of the day and night.

(10) The sign of the zodiac in which the sun is found.

(11) The ancient "unequal" or "planetary" hours, longer in summer than in winter, which divided the period of daylight into 12 equal hours.

Of these types of furniture, the maker's name in an inconspicuous place is always acceptable, as is the year in which the dial was made.

Dial Mottoes. A motto on a sundial has become so common as to be almost expected. Many of them have been used and reused until they seem trite. Collections have been published[3] giving literally hundreds of mottoes which have been either used or suggested for sundials. To be appropriate for this purpose a motto should refer, directly or by inference, to time; and a really apt motto may also incorporate some reference to the dial's location or to its owner or to some other association. An original motto usually has more charm than a "canned" one taken from a standard collection. Thus a dial standing near a babbling brook may fittingly carry the familiar lines of Isaac Watts' hymn:

> Time, like an ever-rolling stream,
> Bears all its sons away.

while a dial on a rocky ledge might bear the lines from Goldsmith's "Deserted Village":

> Self-dependent power can time defy,
> As rocks resist the billows and the sky.

What could be more appropriate than an acrostic of the owner's

[3] One might try for a starter one or more of the following:

Mrs. Alfred Gatty, *The Book of Sun-Dials*, 4th ed., enlarged and re-edited by H. K. F. Eden and Eleanor Lloyd (London: George Bell, 1900).

Charles Leadbetter, *Mechanick Dialling* (London: Caslon, 1773).

Alfred H. Hyatt, *A Book of Sundial Mottoes* (New York: Scott-Thaw, 1903).

Perceval Landon, *Helio-tropes, or New Posies for Sundials* (London: Methuen, 1904).

Alice Morse Earle, *Sun-dials and Roses of Yesterday* (London: Macmillan, 1922).

Launcelot Cross, *The Book of Old Sundials* (London: Foulis, 1914). (This book was illustrated by Warrenton Hogg and is often catalogued under his name.)

name, hopefully more skillfully wrought than this example:

> While now the shadow seems to stay
> As Time enjoys a holiday,
> Unchanging to our simple eye,
> Grave Time, as I can testify,
> Has come—has gone—has passed us by!

Or, on a dial which was a bridal gift:

> Except when clouds
> Disguise the Sun,
> I mark the Hour
> Till Day is done—
> His course is run!

The Equation of Time. A well-informed user of a sundial realizes that the time which it tells must, in most cases, be "corrected" to give the type of time with which we are most familiar, and he will find it convenient to have the necessary information at hand for easy use. If the dial plate is sufficiently large, this can take the form of a table of figures showing the amount by which the dial is "fast" or "slow" on various dates, or one may prefer to make up a small graph of the values of the equation of time from the data of Table A.1 in our Appendix, and transfer this graph to the dial plate as shown in Figure 15.2. Either the table or the graph should combine the longitude correction with the equation of time unless correction for longitude has already been incorporated in the design of the hour lines on the dial.

Showing the Time in Other Cities. It would be easy to construct a sundial to be placed in New York City which showed at any moment what time it was in San Francisco rather than the time in New York. If the dials include longitude corrections the task is especially easy. New York uses Eastern Standard Time, and San Francisco uses Pacific Standard Time. These differ by just 3 hours. When it is 2 P.M. in New York it is 11 A.M. in San Francisco. We construct a dial for New York in the regular way, but after the hour lines have been laid out we merely number them differently. The hour line which would have represented 12 o'clock noon we now label "9," since when it is noon in New York it is 9 A.M. in San Francisco. Correspondingly each other hour line is merely labeled 3 hours earlier than it would be if the dial were to tell New York time.

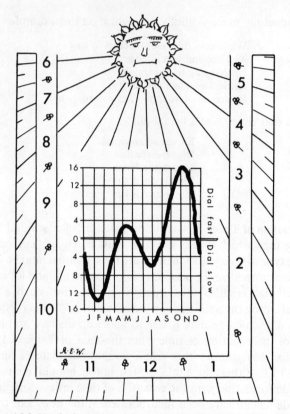

FIGURE 15.2 Vertical sundial with graph showing the equation of time.

Instead of making a dial which shows just the time of some distant city it is likely that we will want to make a dial which shows two kinds of time at once—the time of the place where the dial is located, and the time of some other city as well. Thus the dial of Figure 15.3 was calculated for use in Amherst, Massachusetts (longitude 72°31′ W), and if we read the shadow against the inner scale we will find the L.A.T. in Amherst; but if at the same time we read the shadow against the outer scale it will tell us the corresponding time in Berlin, Germany. Both dials are computed for L.A.T. The longitude of Berlin is 13°25′ E, so the two cities differ in longitude by 85°56′ or approximately 5^h44^m. When it is 8 P.M. in Berlin it is 5^h44^m earlier in Amherst, or 2:16 P.M., and the hour line for 8 P.M. on the outer scale is therefore drawn at the point which corresponds to 2:16 P.M. on

FIGURE 15.3 Horizontal dial showing local apparent time in both
Amherst, Massachusetts, and Berlin, Germany.

the inner scale. All the other hour lines on the outer scale are
placed accordingly.

Still another interesting approach is that illustrated in Figure 15.4,
where the main dial is computed for Farmington, Connecticut.
Farmington keeps Eastern Standard Time and Denver keeps
Mountain Standard Time, 2 hours earlier. When it is noon at Denver
it is 2 P.M. at Farmington, so Denver is shown on the outer border

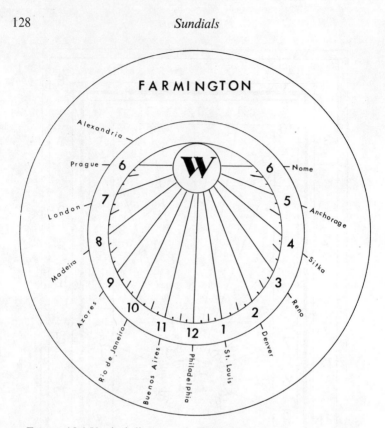

FIGURE 15.4 Vertical direct south dial for Farmington, Connecticut, showing also when it is noon at selected cities around the world.

opposite the hour of 2 P.M. Each of the other cities is placed at the time in Farmington which corresponds to noontime in the distant city; so whenever the shadow, as it circles the inner circle, reaches one of the cities on the outer circle, we know that it is noontime at that distant city.

The Nodus and the Perpendicular Style. Most sundials show the time through a shadow—and really through the straight edge of a shadow which lies among the hour lines. In interpreting dial furniture, however, we do not usually use the entire shadow of the style, but merely some particular point on that shadow cast by a particular selected point on the style called the *nodus*. Figure 15.5 illustrates the terms which we use. The upper part of that figure represents a

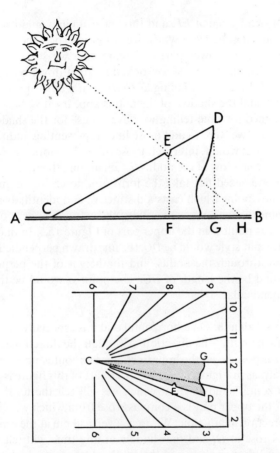

FIGURE 15.5 The style with its nodus (upper) and the shadow which they cast at 1:30 P.M. (lower).

vertical cross section through a horizontal sundial, with the dial plate at *AB* and the gnomon, *CDG*, standing perpendicularly thereon. The angle *GCD* is the *height of the style*, and the line *CD*, which is the upper slanting edge of the gnomon and which casts the shadow used in time-telling is called the *style*. In this case the style is interrupted by a notch at *E* which forms the *nodus*. The line *EF*, drawn through the nodus perpendicular to the dial plate, is called the *height of the perpendicular style*, and the point *F* at the foot of the perpendicular style is called the *foot of the perpendicular style*. As a matter of fact, we could remove all the rest of the gnomon except a slender pin at *EF*, and this vertical pin would itself act as a perpendicular style from which we could interpret most of the dial furniture. In the lower part of Figure 15.5 we see the dial plate itself as it would appear if we looked straight down on it, with the shadow of the gnomon falling at *CGD*, as it would appear at about 1:30 P.M.; and the shadow of the nodus appears at *E*. Since this dial was designed for time-telling we have no use for the shadow of the nodus, but if we were to add more lines representing additional dial furniture, we would interpret those lines by noticing where the point of the shadow of the nodus, *E*, fell among them.

The nodus need not take the form of a notch in the gnomon. It may be any point which casts a distinctive and identifiable shadow. Thus some diallists use as the nodus the extreme upper point of the gnomon, as point *D* in the upper part of Figure 15.5. In that case the perpendicular style would be *DG*, the line drawn perpendicular to the dial plate through the nodus, and the height of the perpendicular style would be the length of *DG*. The point *G* would be the foot of the perpendicular style.

The Sun's Azimuth. On a horizontal dial it is especially easy to show the sun's azimuth—its direction. Although the directional lines can be superimposed over the hour lines, it is less confusing to show them separately, as in Figure 15.6. The lower part of this figure is an ordinary horizontal sundial, from which we will tell the local apparent time. At the upper part is a compass rose from which we will tell the sun's direction at any time. We erect a vertical pin at the center of the compass rose at *O*, and the shadow at any time will tell the sun's direction. When the sun is in the south, at noon, its shadow will fall toward the north, so we label the north point of the compass rose "*S*." Similarly, when the sun is in the southeast, the shadow will lie toward the northwest. For this reason the azimuth points on the compass rose are reversed from the actual directions.

It is also easy to show the sun's direction on a direct south vertical dial. The mariner's compass divides the 360° of the horizon into 32 "points of the compass," each of 11°15'. The 16 points on the north side will lie behind this southern dial and cannot be used. Moreover, when the sun lies due east or west it will be shining parallel to the dial plane, and its shadow will not reach the plane. Table 15.1 shows the remaining compass points, with their angles from the south (azimuths) and the natural tangents of those angles.

FIGURE 15.6 Horizontal sundial with compass rose at the top to indicate the sun's direction.

If we erect a pin as a gnomon, perpendicular to the dial plate, at the point shown in Figure 15.7, the shadow of the outer tip of the pin will move across the lower grid of lines in the course of the day. At noon L.A.T., when the sun is on the meridian, the shadow will fall vertically downward along the line labeled "*S*," but as the afternoon

TABLE 15.1
AZIMUTHS FOR A VERTICAL DIRECT SOUTH
DIAL AND THEIR TANGENTS

compass direction	azimuth from the south	tangent of the azimuth
S	0°00′	0.000
S^{b}E \ S^{b}W	11°15′	0.199
SEE \ SSW	22°30′	0.414
SE^{b}S \ SW^{b}S	33°45′	0.668
SE \ SW	45°00′	1.000
SE^{b}E \ SW^{b}W	56°15′	1.497
ESE \ WSW	67°30′	2.414
E^{b}S \ W^{b}S	78°45′	5.027
E \ W	90°00′	∞

wears along and the sun moves toward the west, the shadow will move toward the east across the remaining lines. Every time it moves one "point of the compass" it will have moved 11°15′. Thus, when the sun's direction is just one point before noon (south by east, represented by S^{b}E), and again when it is just one point after noon (S^{b}W), its angle from the gnomon will be 11°15′, and the distance of the tip of the shadow from the vertical south line will be the length of the pin gnomon multiplied by the tangent of the angle. If we use the length of the gnomon as 1.000, then the distances of the various lines on the lower grid from the central south line will always be equal to the tangent of the angle as shown in the table above—0.414 times the height of the pin when the sun is SSE or SSW, 1.000 times the height of the pin when the sun is SE or SW, and so forth. We would usually select a pin of such length as to keep the tip of the shadow on the dial plate for a good portion of the day, as was done in Figure 15.7. The length of the little pin will always be exactly equal to the distance on the lower grid from the south line to either the SE or the SW line.

If our vertical dial declines toward the east or the west, the azimuth lines will still form a grid of vertical parallel lines, but displaced so

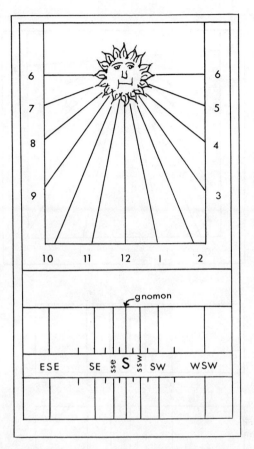

FIGURE 15.7 Ordinary vertical direct south dial at the top, and underneath a grid showing the direction of the sun at any time.

that the south line lies to the left of center if the dial declines toward the west, or to the right of center if the dial declines toward the east. For example, let us consider a vertical dial declining S 20° E in latitude 40° N. The upper half of Figure 15.8 shows the ordinary dial designed in accordance with the principles of Chapter 10. We could place a nodus on the gnomon of this dial and superimpose a grid of vertical azimuth lines on the radiating set of hour lines, but again, to avoid confusion, we here show the azimuth grid as a separate dial at the bottom, with its own gnomon in the form of a pin set at the center perpendicular to the dial plate.

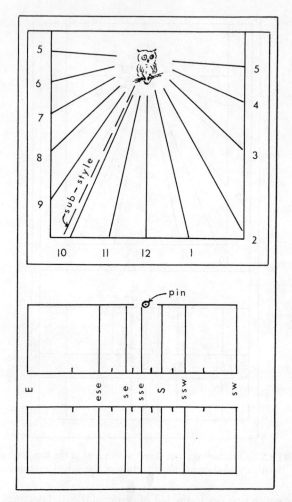

FIGURE 15.8 Vertical declining dial (top) with scale at bottom showing the direction of the sun.

Since this dial declines S 20° E, the sun will be directly in front of the dial when the sun itself is 20° east of south, and at this moment the shadow of the pin will fall vertically down the lower grid of azimuth lines. If we refer back to Table 15.1 we will note that S 20° E lies between the compass points of S♭E and SSE, and closer to the latter. In fact, SSE is 22°30′ east of south, or just 2°30′ farther east than the direction the dial faces. In other words, the sun will lie in the direction SSE 2°30′ before it is perpendicular to the face of the dial, and the grid line for the direction SSE should lie to the left of the gnomon by a distance equal to the tangent of 2°30′, or 0.044 times the height of the pin. Taking one other azimuth for purposes of illustration, we note from Table 15.1 that the azimuth SW is 45° west of south. But our dial declines 20° east of south, so the direction SW is 45° + 20° or 65° west of the dial. And since the tangent of 65° is 2.145, we draw the grid line for azimuth SW 2.145 times the height of the gnomon to the right of the pin. Computation will show that for this dial the major compass points lie at the following angles from the direction of the dial:

azimuth	E	ESE	SE	SSE	S	SSW	SW	WSW
angle	70°	$47\frac{1}{2}°$	25°	$2\frac{1}{2}°$	20°	$42\frac{1}{2}°$	65°	$87\frac{1}{2}°$

The distances of the corresponding azimuth lines from the vertical line under the gnomon pin will be equal to the product of the height of the pin times the tangent of the given angle.

The Sun's Altitude. Old sundials often included lines or scales showing the sun's altitude, or angular distance above the horizon. On a horizontal dial the problem is simple, since any given altitude will be represented by a circle centered at the foot of the perpendicular style with a radius equal to the product of the height of the perpendicular style and the cotangent of the sun's altitude. Thus, if the height of the perpendicular style is 2.5 inches and we wish to know when the sun's altitude is 50°, we draw a circle centered at the foot of the perpendicular style with a radius of 2.5 times cot 50°, or 2.5 × 0.839 or 2.0975 inches. Whenever the shadow of the nodus touches this circle, the sun has an altitude of 50°. Altitudes below 10°–15° are seldom shown, since their circles would be too large. The maximum altitude which the sun can reach in any latitude can be found by subtracting that latitude from 113°26.7′. Thus the sun's maximum altitude at latitude 40° (which occurs at noon on June 21) is 113°26.7′ − 40° = 73°26.7′. Figure 15.9 shows a horizontal dial computed for latitude 40° with the usual straight hour lines radiating from the dial center, and superimposed thereon 6 circles showing altitudes from 20° to 70°. Also shown is the shadow of the gnomon as it would appear at about 9:30 A.M., with the shadow of the nodus indicating that the sun's altitude is 40°. In this case, as in Figures 15.6, 15.7, and 15.8, the altitudes could have been shown on a separate scale rather

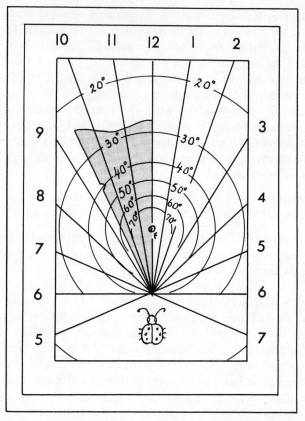

FIGURE 15.9 Horizontal dial showing the local apparent time (9 : 30 A.M.) and the sun's altitude (40°). The foot of the perpendicular style is at the little circle marked *F*.

than superimposed on the hour lines, and the altitude indicated by a separate vertical pin rather than by the nodus on the gnomon of the time dial.

Old-time dial makers, instead of showing solar altitudes of 20°, 30°, etc., often inserted circles to show when any vertical object would cast a shadow equal to its own height, twice its own height, three times its own height, and so forth. Such circles appear on the old dial of Figure 15.1, where they are labeled "Aequalis," "Dupla," "Tripla," "Quadrupla," and "Quintupla." The radii of these circles are, of course, equal to one, two, three, four, and five times the height of the perpendicular style.

Lines of Declination. As the sun moves across the sky from east to west each day, the shadow which is cast on a horizontal surface by the upper tip of a vertical pin (or by the nodus on the slanting style of a horizontal sundial) traces a path from west to east. At the times of the equinoxes this path is a straight line running due east and west perpendicular to the meridian or to the 12-o'clock line. At all other times of year the path is curved, being part of an hyperbola in our latitudes. When the sun is north of the equator during the summer months this path is concave toward the vertical pin which casts the shadow, but during the remainder of the year it is convex toward the pin. The three situations are apparent from Figure 15.10, which

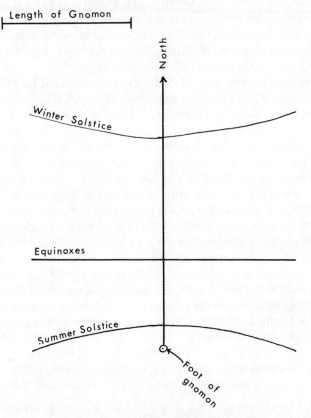

FIGURE 15.10 Paths traced by the shadow of the nodus at the times of the equinoxes and solstices in latitude 34° N. The gnomon is a vertical pin of the length shown in the insert at the top.

shows the paths which the shadow would trace at latitude 34° N if the vertical pin had a height equal to the length of the line at the top of the figure. While the lines are shown for but four days of the year (two solstices and two equinoxes), there would be a different but distinctive path on any other date, depending solely on the declination of the sun. These other lines would all lie within the limits of those shown in Figure 15.10. If we know the sun's declination on any day we can draw the line of declination for that day, so that our dial will serve as a calendar to tell the day of the year as well as a timepiece to tell the time of day. In fact, a leading expert on early scientific instruments has pointed out[4] that the hour lines of those old dials which have survived are often not numbered even though the lines of declination are frequently identified. This led him to conclude that "one cannot well avoid the suggestion that the chief function of the sundial in antiquity was not the determination of the time of day, but season of year."

If one does include lines of declination on a sundial he usually confines himself to the three lines shown in Figure 15.10, showing the dates of the solstices and equinoxes. Those whose interest in the zodiac runs deeper may wish to add lines showing the sun's entrance into the other "signs," as was done in the old dial shown in Figure 15.1. The dates and corresponding solar declinations for these entrances appear in Table A.3 of our Appendix. But we need not confine ourselves to lines with astronomical significance. For example, a dial given as a wedding present might fittingly carry the line of declination along which the shadow would pass each successive year on the wedding anniversary; or lines might mark birthdays of members of the family, or other events which one wishes to commemorate. But it is easy to carry the matter too far and confuse the dial plate with a multiplicity of lines. Since the method of drawing all lines of declination is the same, we shall illustrate with the lines shown in Figure 15.10 and allow the reader to apply the method to other dates at his pleasure.

The Use of Tables. If we have access to the appropriate tables,[5] the process is greatly simplified. We must know our latitude, the sun's

---

[4] Derek J. DeSolla Price, *Vistas in Astronomy* Vol. 9, p. 41. (Oxford: Pergamon Press, 1968.) See a similar reference to him in the footnote of page 5.

[5] Hydrographic Office *Tables of Computed Altitude and Azimuth*, see footnote, page 90.

declination on the date for which the line is to be drawn, and the height of the perpendicular style. The tables will then tell us the sun's altitude at any time of day, and the cotangent of the altitude multiplied by the height of the perpendicular style will give us the distance from the foot of the perpendicular style to the tip of the shadow—and this tip of the shadow, in turn, must lie on the appropriate hour line. Let us take an example. The tables tell us that on March 20, when the sun has its maximum northerly declination of 23°30′, at 2 P.M. in latitude 40° N the sun's altitude is 59°50.9′. The cotangent of this angle is 0.581. If the height of the perpendicular style is 1.5 inches, the length of the shadow will be (1.5)(0.581) = 0.872 inches. If we measure out from the foot of the perpendicular style (*not* from the dial center) to the point on the hour line of 2 P.M. which is 0.872 inches away, we will have one point on the line of declination for March 20. We find similarly where this line of declination crosses each of the other hour lines, and draw a smooth curve through the points. This smooth curve is our desired line of declination.

If we do not have the tables handy, we can compute the sun's altitude and azimuth at any moment of time for any latitude and solar declination. This requires the solution of three formulas, as follows:

(1)	$\tan M = (\tan \text{dec})/(\cos t)$
(2)	$\tan Z = (\cos M)(\tan t)/\sin (\phi - M)$
(3)	$\tan h = (\cos Z)[\cot (\phi - M)]$

In these formulas M is merely an intermediate value for use in the subsequent equations, dec is the sun's declination, t is the sun's hour angle, ϕ is the latitude, Z is the sun's azimuth from the south, and h is the sun's altitude. If we solve these equations for the case where dec $= +23\frac{1}{2}°$ (as on March 20), $t = 30°$ (as at 10 A.M. or 2 P.M.), and $\phi = 40°$ N, we get:

$$M = 26°39.6′; \qquad Z = 65°54.4′; \qquad h = 59°51.0′.$$

This value of solar altitude differs by but one-tenth of a minute from that taken from the tables. We would similarly solve these equations for the other hours of the day and get the points of intersection on the other hour lines through which we would draw a smooth curve as the line of declination.

Graphic Approach. For those who dislike even the slight mathematics involved in the preceding steps, one can find the lines of solar declination entirely by graphic means. Our approach follows that of a famous writer on dialling of three centuries ago.[6] We illustrate with Figure 15.11, which is in two parts. The upper part, which we shall refer to as part A, is an ordinary horizontal sundial for latitude 33° N. The lower part, which we shall call part B, supplements it.

(1) In part B, draw triangle OST of any convenient size so that:

    (a) ST is the desired height of the perpendicular style.

    (b) S is the foot of the perpendicular style.

    (c) T is the nodus, and angle OST is a right angle.

    (d) OS is the sub-style with O the dial center.

    (e) OT is the style and angle SOT equals the latitude.

The size of this triangle will determine the scale of the final diagram.

(2) In part A, lay off along the 12-o'clock line from the dial center at O the distance OS equal to OS in part B. Then S in part A will be the foot of the perpendicular style which will have a height equal to ST of part B.

(3) In part B, draw TE perpendicular to OT. This will represent the equator.

(4) In part B, draw TD and TF making angles DTE and ETF each equal to the sun's maximum declination, $23\frac{1}{2}°$.

    (a) TD represents the Tropic of Cancer and TF the Tropic of Capricorn. If we wanted to draw some other line of declination we would lay off an angle from ET equal to that declination—on the side nearer to O if the declination were northern and on the side away from O if it were southern.

(5) In part B, extend OS to intersect TD, TE and TF. Mark its intersection with TE with "a."

We now work back and forth between parts A and B:

    (a) Take distance Oa from part B and transfer it to part A from O along the 12-o'clock line to "a."

    (b) In part A draw a line through a perpendicular to the 12-o'clock line. This will be the equator, and the shadow of the nodus will fall on this line on the days of the equinoxes. Label the intersections of the equator with the various hour lines of part A with b, c and d. (Intersections with other hour lines would require that these other lines be extended beyond the present diagram.)

[6] William Leybourn, *Dialling* (London: 1682).

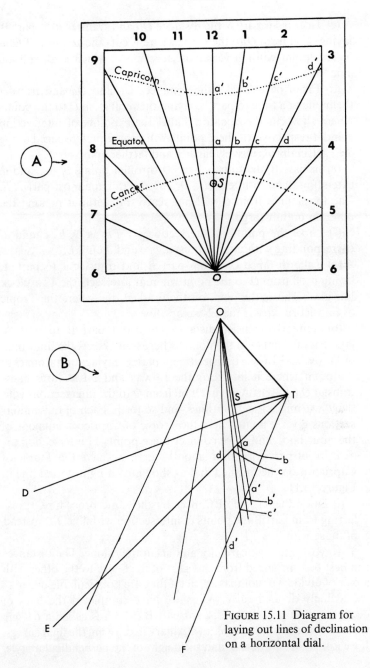

FIGURE 15.11 Diagram for laying out lines of declination on a horizontal dial.

(c) Take from part A the distance Ob and transfer it to part B, laying it off from O to the point where it cuts the equator, TE, at b. This point will fall very close to point a, which is already on line TE.

(d) From part A take the distances Oc and Od and transfer them one at a time to part B, laying them off from O to the points where they cut the equator. Label these points of intersection c and d respectively. These points will be unequally spaced along TE, getting progressively farther and farther apart.

(e) In part B, draw lines from O through points a, b, c and d, extending them to intersect TD, TE and (for the most part) TF. The lower lines from O will not reach TF without making the diagram unduly large.

(f) Label the points of intersection on TF as a', b', c' and d' corresponding with intersections a, b, c and d on TE.

(g) Take distance Oa' from part B and transfer it to part A, laying it off from O to the point where it intersects the 12-o'clock line at a'. This is the point on the finished dial where the Tropic of Capricorn crosses the 12-o'clock line.

(h) Similarly transfer distance Ob' from part B to part A, laying it off from O to the points where it intersects the hour lines of 11 A.M. and 1 P.M.; transfer Oc' to part A, laying it off from O to the point where it intersects the 10 A.M. and 2 P.M. hour lines; transfer Od' to part A, laying it off from O to the intersections with the 9 A.M. and 3 P.M. hour lines, and so forth. Each of these intersections gives a point where the Tropic of Capricorn cuts one of the hour lines, and if we connect these points of intersection (a', b', c', d' of part A) with a smooth curve we have the Tropic of Capricorn on the finished dial as shown by a light dotted line in Figure 15.11.

(i) Draw the Tropic of Cancer by the same procedure, transferring from part B the points of intersection with line TD instead of those with line TF.

This work must be carefully and accurately done. The distances can best be transferred from one part of the figure to the other with a pair of dividers or compasses, but if they are not available one may use a finely divided ruler. When we have completed the lines of declination on part A, we discard part B (which is called a *trigon*), since its sole use is to serve as an auxiliary in drawing the final curves. If we know the latitude, ϕ, and the height of the perpendicular style, *h*, we can find the distance from the dial center to the foot of the perpendicular style as $(h)(\cot \phi)$; and the distance from the dial center to the intersection of the equator and the 12-o'clock line as $(h)/(\sin \phi)(\cos \phi)$. At this latter point the equator will cross the 12-o'clock line at right angles.

Vertical Direct South Dials. As we have learned in Chapter 6, every vertical direct south dial is a duplicate of the horizontal dial computed for the colatitude. Consequently if we want to find the lines of declination on a vertical direct south dial at latitude 57° N, we merely lay out a horizontal dial for latitude 33° N and add the lines of declination to it by the rules which we have just outlined. But Figure 15.11 and its accompanying material already cover a horizontal dial for latitude 33° N. Our work is already done. Of course, we must invert the dial shown in Figure 15.11 and number the hour lines counterclockwise rather than clockwise—and after we trace off the Equator and the two Tropics we also reverse their order, putting the Tropic of Capricorn at the top close to the foot of the perpendicular style and the Tropic of Cancer at the bottom. But everything else stays exactly as it was in Figure 15.11. We use the same position for the foot of the perpendicular style, and the same height of the perpendicular style. Figure 15.12 shows the vertical direct south dial with its lines of declination and, in fact, Figure 15.12 was merely traced from Figure 15.11 with the necessary inversions. We must keep in mind, of course, that Figures 15.11 and 15.12 are for dials in different latitudes. Figure 15.11 shows a horizontal dial for latitude 33° N. Had we wanted a direct vertical south dial for this same latitude we would have started by laying out a horizontal dial for latitude 57° N, after which we would have found the lines of declination for that horizontal dial.

Polar Dials and Vertical Direct East or West Dials. The drawing of lines of declination on these dials is unusually easy. We have the equatorial line already drawn down the center of the dial (see Figures 8.1, 8.2, and 9.1) and the hour lines in place. The distances between the hour lines depend on the height of the perpendicular style, being equal to this height multiplied by the tangent of the sun's hour angle from 6 A.M. (direct east dial), 6 P.M. (direct west dial), or 12 noon (polar dial). We now merely measure out along each hour line the proper distance to reach the lines of declination—and

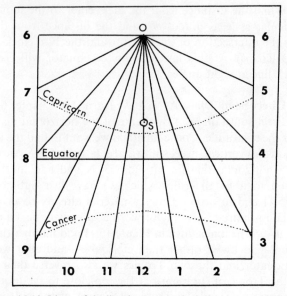

FIGURE 15.12 Lines of declination on vertical direct south dial, based on the dial of figure 15.11.

for the Tropics of Cancer and Capricorn these distances are found by multiplying the height of the style by the factor shown in Table 15.2 for each hour line (counting the hour lines from 6 A.M., 6 P.M., or 12 noon as mentioned above).

TABLE 15.2
FACTORS FOR FINDING DISTANCES TO THE LINES OF DECLINATION AT THE TROPICS. (VERTICAL EAST OR WEST, OR POLAR DIALS)

hour	factor	hour	factor
0	0.434	3	0.613
$0\frac{1}{2}$	0.437	$3\frac{1}{2}$	0.712
1	0.449	4	0.867
$1\frac{1}{2}$	0.470	$4\frac{1}{2}$	1.133
2	0.501	5	1.675
$2\frac{1}{2}$	0.547	$5\frac{1}{2}$	3.322

For example, Figure 15.13 shows the equator and the hour lines of a polar dial laid out in accordance with the rules covered in Chapter 9. The height of the perpendicular style in such a dial is equal to the distance from the 12-o'clock to the 3-o'clock hour line. We start at the point where the equator and the 12-o'clock lines intersect and

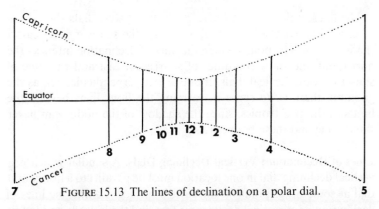

FIGURE 15.13 The lines of declination on a polar dial.

measure out in each direction along the 12-o'clock hour line distances of 0.434 times the height of the perpendicular style (which is 0.434 times the distance between the hour lines of 12 and 3 o'clock). We then move either way one hour to the hour lines of 11 A.M. and 1 P.M., and measure out from the equator along these hour lines in either direction distances of 0.449 times the height of the perpendicular style. On the hour lines for 10 A.M. and 2 P.M. we measure out 0.501 times the height, and so forth, using the figures in the preceding table. This gives us two points on each hour line, one above and one below the equator. We connect them with smooth curves, shown as light dotted lines in Figure 15.13, for the lines of declination on the days of the solstices. If, at the junction of the equator and the 12-o'clock line, we erect a vertical pin with a height equal to the distance between the 12-o'clock and the 3-o'clock lines the shadow of the tip of this pin will fall on the Tropic of Capricorn on December 21 each year, and on the Tropic of Cancer on June 21. On other days it will fall between these two limiting lines, and on the days of the equinoxes it will fall along the equator at the center of the dial. If our gnomon takes the form shown in Figure 8.3 we can file a notch in the center of the style to serve as a nodus, and the shadow of this nodus will indicate the date. We can, of course, show lines of declination for other dates if we wish.[7]

[7] If d is the sun's declination on any given day and if t is the sun's hour angle measured from 12-o'clock on polar dials, from 6 A.M. on vertical east dials and from 6 P.M., on vertical west dials, we find points on the line of declination corresponding to any given solar declination as follows: Measure from the 12-o'clock line along the equator (or from the appropriate 6-o'clock line) a distance equal to tan t. There erect a line perpendicular to the equator and measure out from the equator along this line a distance equal to tan d/cos t. This will be a point on the desired line of declination.

The lines of declination on vertical east or west dials should not be carried above the horizontal line, since the sun can never cast a shadow there. The points where the lines of declination intersect the horizontal line show the times of sunrise on east and the time of sunset on west vertical dials. If one uses a perpendicular pin as the gnomon, there is no sense in carrying the hour lines beyond the limits of the two tropics, since the shadow of the nodus can never wander beyond these boundaries.

Lines of Declination: Vertical Declining Dials. A sundial which is a vertical declining dial in one location must lie parallel to a horizontal dial at some other point on the earth's surface. To find the lines of declination on a vertical decliner we first find the latitude and longitude of that other spot where the dial would be horizontal, construct an ordinary horizontal dial for that location, and add to it the lines of declination as with any other horizontal dial.

Suppose, for example, that we want to find the lines of declination on a vertical dial declining S 19°30′ E in latitude 41°42.9′ N and longitude 72°49.0′ W. Using the methods of Chapter 10 we find the following values for this dial:

$$SD = 20.5° \qquad SH = 44.7° \qquad DL = 28.0°.$$

The style height, 44.7°, is the new latitude in which our dial would be horizontal. The difference in longitude tells us that the dial would be horizontal at a longitude 28.0° farther east (since the declination is east), or at longitude 44.8° W. So we draw an ordinary horizontal dial for latitude 44.7°, and add the lines of declination for such a dial. These lines of declination will be symmetrical about the 12-o'clock line of that dial. But we learned in Chapter 10 that in a vertical decliner the sub-style (which is the 12-o'clock line of our horizontal dial) stands at an angle to the 12-o'clock line, in this case an angle of 20.5°. So we draw the hour lines for our actual vertical declining dial by the methods of Chapter 10, and superimpose on them the lines of declination from our corresponding horizontal dial. This will give the final pattern illustrated in Figure 15.14. In this figure everything is just as it would be had we drawn the dial by the rules of Chapter 10 except that we have added the lines of declination for a horizontal dial in latitude 44.7° and have centered them around the sub-style. We have also added the horizontal line, which passes through the foot of the perpendicular style and will actually be

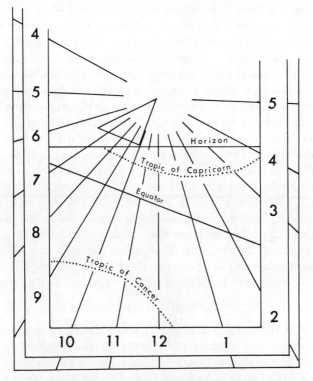

FIGURE 15.14 Vertical dial declining S 19° 30′ E in Farmington, Connecticut; latitude 41° 42.9′ N and longitude 72° 49.0′ W; with the Equator and tropics.

horizontal when our finished dial is set in place. The effect of the dial's declination is to "tip" the lines of declination, or to skew them around, centered at the foot of the perpendicular style, with the Equator perpendicular to the sub-style rather than to the 12-o'clock line of the vertical dial. The Tropic of Capricorn is not carried above the horizontal line in the finished dial. The fact that this tropic crosses the horizon at about 7:30 A.M. indicates that this is the time of sunrise for the latitude of this dial on December 21, when the sun is at the Tropic of Capricorn. If we were to extend the diagram slightly farther to the left we would find that the Equator would cross the horizon just at the 6-o'clock line, indicating that sunrise occurs at 6 A.M. when the sun is on the celestial equator.

The Length of Day. The *horizontal line* on a dial plate is the line formed by the intersection of the dial plate and a horizontal plane passed through the nodus. (There is no horizontal line on a horizontal dial since a plane passed horizontally through the nodus is parallel to the dial plate and never intersects it.) We have noted that one can tell the time of sunrise on a vertical direct east plane, and the time of sunset on a vertical direct west plane; and from these times one can determine the length of the day. More commonly, however, the "lines of the length of day" on old dials were lines of solar declination which corresponded to various lengths of day. For example, one discovers what the solar declination must be in our latitude for the day's length to be 10 hours, 11 hours, 12 hours, and so forth, and then draws the lines of declination for these solar declinations. If, now, the shadow of the nodus falls on the 10-hour "length of day" line, we know that the day is 10 hours long. The 12-hour length is easy, since everywhere the days are 12 hours long on the days of the equinoxes. The Equator on any dial is the line of 12-hour days. For days of other lengths we use the formula:

$$\log \tan d = \log \sin (L/2) + \log \cot \phi,$$

where d is the sun's declination, L is the difference between the length of day and 12 hours, and ϕ is the latitude. Suppose we wish to show on the dial of Figure 15.12 the line of declination which shows a day's length of 14 hours. Since L is the difference between the length of day and 12 hours, L is 2 hours or 30°. Figure 15.12 shows a dial designed for use in latitude 57° N, so $\phi = 57°$ and $L/2 = 15°$. Substituting these in the formula gives a value of d of 9°32.5′. When the sun has a declination of 9°32.5′ north (as on about April 15 and August 29) the days will be 14 hours long in latitude 57° N—and when the solar declination is −9°32.5′ (as on about February 24 and October 19) the days will be 2 hours less than 12 hours or 10 hours long. If we draw lines of declination for solar declinations of +9°32.5′ and −9°32.5′, these will be the lines for lengths of day of 14 hours and of 10 hours.

Signs of the Zodiac. Our ancestors appear to have had more interest in the signs of the zodiac than most modern people have, and it was not uncommon to show on a sundial the lines of solar declination along which the shadow of the nodus would move on the days when the sun was entering the various "signs." The declination which the sun has when it enters each sign is shown in Table A.3 of our Appendix, and one merely lays out on the face of the dial the line of declination corresponding to the sign, and labels it with the symbol of the sign, as shown in the old dial of Figure 15.1. If the shadow of the nodus falls on one of these lines we know that the sun is that day entering the corresponding sign.

16
Portable Dials

> And then he drew a dial from his poke,
> And looking at it with lack-lustre eye,
> Says, very wisely, "It is ten o'clock;
> Thus may we see," quoth he, "how the world wags."
> —Shakespeare, *As You Like It*, II, vii.

Prior to the early seventeenth century pocket watches were uncommon, expensive, and unreliable, and the traveler who wished to keep track of the time was forced to rely on a portable sundial. We are told that King Charles I carried a silver pocket sundial, and that on the evening preceding his execution he entrusted it to his attendant as a last gift to his son, the Duke of York. Lafayette gave a silver pocket dial to George Washington, and several dials in modern collections were carried by officers during the American Revolution. There were also various types of inexpensive portable dials in common use by common people.

The Shepherd's Dial. Among the simplest and most widely used of the portable dials was one which was variously called the shepherd's dial, the pillar dial, the traveler's dial, the chilindre, and the cylinder. It was not very accurate, but was easily made and inexpensive. We recall that the "gentil monk" of the Canterbury Tales, inviting the "gode wyf" to dinner, says,

> "Goth now your wey," quod he, "al stille and softe,
> And lat us dyne as sone as that ye may;
> For by my chilindre it is pryme of day."

The shepherd's dial indicates the time of day from the sun's altitude, which depends not only on the time of day, but also on the latitude and time of year. Consequently the shepherd's dial is designed for a particular latitude and is adjustable for the date. As its name implies, it is cylindrical in form, and the hour lines appear as curves on its rounded face. (See Figure 16.1.) In use the dial is hung by a string with the gnomon (a horizontal pin at the top) extended toward the sun with its shadow falling among the hour lines below. The top of the dial rotates, and the pin is set over that part of the scale corresponding with the day of the year. The cylinder is usually hollow to contain the gnomon when it is not in use.

The hour lines are either drawn on a paper which is glued to the cylinder or inscribed directly on the surface of the cylinder. As a preliminary to construction, we make up a table showing for selected days of the year the sun's altitude at each hour of the day in the latitude where the dial is to be used. Methods of computing these altitudes were given in the preceding chapter, or one can look them up in published tables.[1] The finished chart of the hour lines will look like Figure 16.2, and our first step must be to determine the length and height of this chart for our own case. Since the chart is to be wound around a cylinder, its length must be equal to or slightly less

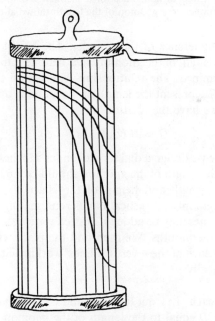

FIGURE 16.1 The shepherd's dial.

[1] See page 90.

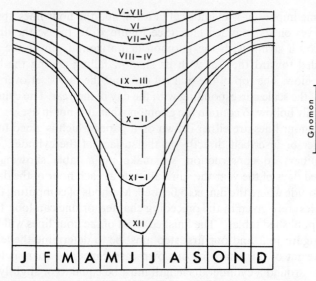

FIGURE 16.2 Chart of hour lines for a shepherd's dial at 42° latitude, proportioned to a gnomon of the length shown at the right.

than the circumference of the cylinder. The height of the chart is related to the length of the gnomon. If either one is fixed, the other is thereby determined. The relationship depends on the latitude, ϕ; and if we let G represent the length of the gnomon and H the height of the chart we have the relationships:

$$G = H \cot (113\tfrac{1}{2}° - \phi).$$

If, for example we design a dial for a cylinder 5 inches high for use at latitude 42°, the length of the gnomon should be 1.67 inches.

Having the length and height of the chart of Figure 16.2, we divide the length into 12 vertical bands representing the 12 months. Intermediate lines may be added if desired to represent the 10th and 20th days of the months. We then find the place where each hour line intersects each of these vertical lines, working either graphically or mathematically.

Graphic Approach. Lay out lines AD and CE of Figure 16.3 at right angles, with CD equal to the length of the gnomon. With D as the center and with any convenient radius draw the circular arc BF and divide it into 10° arcs. From D draw straight lines to the divisions

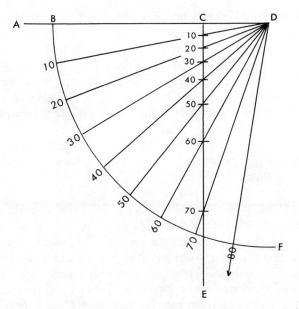

FIGURE 16.3 Developing a scale for determining the locations of the
hour lines on a shepherd's dial. *CD* is the height of the gnomon and *CE*
is the scale.

on *BF*, and the places where these lines intersect *CE* will form a scale
for drawing the hour lines. For any given day of the year, say June 1,
we look up in our table the sun's altitude at each hour of that day,
and lay our scale (*CE*) along the vertical line corresponding to that
date with *C* at the top. We mark the point on the vertical line corresponding to the scale division showing the sun's altitude at each hour
of the day, continuing this for the other days, and finally connecting
the points for each hour to give the hour lines themselves as shown
in Figure 16.2.

Arithmetic Approach. If we prefer we can draw the hour lines by
computation. Given the sun's altitude, *A*, at a given hour on a
selected day, we compute the distance, *D*, of the hour line from the
top of the chart as follows:

$$D = (\tan A)(G),$$

where *G* is the length of the gnomon. For example, in latitude 42° N
at 2 P.M. on November 1 the sun's altitude is 26.2°. Our formula

tells us that the 2-o'clock hour line should intersect the November 1 line 1.476 inches down from the top of the chart if the gnomon is 3 inches long. We repeat the operation to find where each of the hour lines crosses each of the date lines, and connect all the points for any hour to form the hour line itself. Since the sun's altitude will be the same at 10 A.M. and 2 P.M. L.A.T., we draw a single line for these two hours—and for any other two hours which are equidistant from noon.

The hour lines on a shepherd's dial lie close together in winter, and also near noontime, so that the instrument is least accurate at those dates and times.

Tablet Dials. A tablet dial is merely two small distinct sundials hinged together with a common gnomon. One is an ordinary horizontal dial and the other a vertical direct south dial, both in miniature size so that, when folded at the hinge, they can fit into the pocket. The dials will appear as shown in Figure 16.4, where the common gnomon is a string stretched from a point near the top of the vertical dial to a point near the outer edge of the horizontal dial. The string casts its shadow across both dials, and the time is read from either dial.

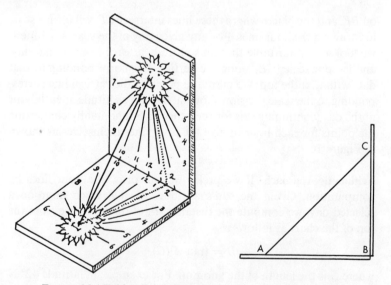

FIGURE 16.4 Tablet dial (left) and side view of the same (right).

The string gnomon must be of such a length that when it is stretched tight the two dials will stand at right angles, and in order that it may have the proper slope for the latitude the two dials must be properly proportioned one to the other. The relationship between the dials is best explained by reference to the illustration on the right of Figure 16.4, where we see the dials in side view with the string gnomon running from *A* to *C*. In order that the string may have the proper slope for the latitude, ϕ, the horizontal distance, *H*, from *A* to *B* and the vertical distance, *V*, from *C* to *B*, must bear the relationship:

$$H = V \cot \phi.$$

There is an unfortunate misconception that if a tablet dial is turned until both dials show the same time, the dials will be properly oriented. Nothing could be further from the truth. If we look again at Figure 16.4 we note that the shadow falls between the hours of 1 P.M. and 2 P.M. on both dials. If we were to twist the dials until the shadow fell between, say, 10 A.M. and 11 A.M. on the vertical dial, it would also have to fall between 10 A.M. and 11 A.M. on the horizontal dial. No matter how we twist and turn the dials, if the shadow of the string falls on either dial, it will fall on both, and will show the same time on both. For proper orientation these dials often carried a tiny compass embedded in the horizontal dial and a tiny plumb line attached to the vertical dial to assure its verticality.

Cubic Dials. While we are treating of instruments which combine two or more separate sundials, we should mention the little cubic dials which were not uncommon three or four centuries ago. Although their shape was such that they could hardly be classed as "pocket" dials, they were small enough to be portable. The typical cubic dial, depicted at the center of Figure 16.5, carried five standard sundials—a horizontal dial on the upper face, and vertical direct north, south, east and west dials on the side faces. The four gnomons, all pointing toward the celestial pole, have parallel styles. The sun shines on two or three faces of the cube at a time, and all faces show the same time no matter how the instrument is oriented. A small compass in the base is used for orientation.

The five dials of the cubic dial are all drawn for the same latitude, but the instrument is made "universal" (usable in all latitudes) by placing a joint in the pedestal on which the cube stands. If we want to use the dial in some latitude other than that for which it was designed

FIGURE 16.5 Three small portable dials from the author's collection.

we merely tilt the cube until all the gnomons point toward the celestial pole, thus making the correction described in Chapter 5 for adjusting a dial to a new latitude.

Portable Cross Dials. Another form of portable sundial sometimes found in collections takes the form of a small cross, often of silver or brass, with the hour lines engraved on the cross arms. The cross was attached to its base with a pivot, so that it could be set up with the main axis parallel to the plane of the Equator, and as the sun moved across the sky the shadow of the main upright moved across the hour lines on the cross arms. These dials were miniatures of the large memorial cross dials described in detail in Chapter 18. Since full directions for laying them out are given there, we need not repeat them here.

Pocket Equatorial Dials. During the sixteenth to eighteenth centuries diminutive folding pocket dials were popular, taking the form, when opened, of an equatorial ring carrying the hour lines and a pin gnomon standing perpendicular to the ring at its center. Such a dial, made in Augsberg, Germany, about 1800, appears at the left in Figure 16.5. The semi-circular arc carries a scale by means of which the ring may be adjusted for the latitude, and a compass in the base is used for orientation. When folded, these dials were little larger than a modern pocket watch, and could be relied upon to give fairly accurate readings of the time whenever the sun shone. The pin which served as a gnomon was pivoted so it could be turned below the plane of the equatorial ring to show the time during the winter months when the sun is below the celestial equator.

The Universal Ring Dial. This type of portable dial, also called the "astronomical ring dial," is very ancient, yet it remained in use until well into the eighteenth century. It is the type of dial, mentioned in the opening paragraph of this chapter, which Lafayette gave to George Washington. It is called "universal" because it is adjustable for use in any latitude. It is really a miniature folding armillary sphere, the theory of which is covered in the following chapter.

The construction of the universal ring dial can best be understood by reference to Figure 16.5, where it appears at the right. Two flat rings, 4–5 inches in diameter, are folded together for storage or transit—one of the rings being just enough smaller than the other so that they nest together when folded. When the instrument is in use they are unfolded until they are mutually perpendicular, as shown in the figure. The dial is suspended by a small ring at the top, which can be moved along the outer circle to the point which corresponds with the latitude. Pivoted across the center of the outer ring is a thin metal "bridge" which can be seen in the figure sloping downward from left to right. The bridge is slotted and bears a cursor pierced with a tiny hole. The cursor is moved to that point on the bridge which corresponds to the sun's declination on the day of the observation. The sun shining through the tiny hole casts a beam of light on the inner surface of the time ring, which is calibrated to show the hours of the day. The three dials of Figure 16.5 are all shown as though viewed from the west, with the gnomons slanting upward to the left toward the celestial pole.

Perforated Ring Dials. Dials were sometimes made in the form of a rather broad circular band pierced in the middle of one side with a tiny hole through which a sunbeam was projected among hour lines inscribed on the opposite interior surface of the ring. In use, the ring was hung in a vertical position with the aperture turned toward the sun as in Figure 16.6. The time was told by the sun's altitude, and hence the hour lines took the general shape of those found on a shepherd's dial described earlier in this chapter.

The method of drawing the hour lines for these dials is illustrated in Figure 16.7. The circle in the upper part of that figure represents the ring, hung vertically from M. AOB is a horizontal line through

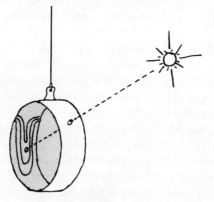

Figure 16.6 Perforated ring dial.

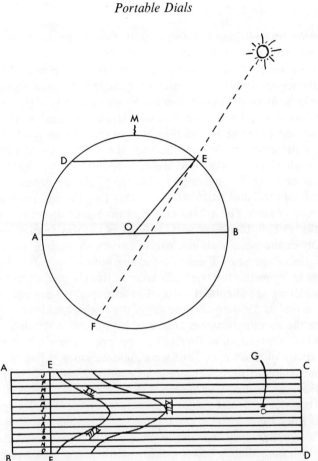

FIGURE 16.7 Schematic view of a perforated sun ring (upper), and inside surface of the ring when straightened out (below).

the center of the ring, so arcs *AM* and *BM* are each 90°. The aperture is at *E*, which can be placed at any point on the upper half of the circle, but is best so placed that angle *BOE* is equal to the colatitude. *ED* is parallel to *AOB* and is consequently horizontal. The sun casts a beam through *E* to *F* when the sun's altitude is equal to angle *DEF*; but since an angle inscribed in a circle is measured by half the subtended arc, the arc *DF* will be equal to twice the sun's altitude. At sunrise and sunset, when the sun's altitude is zero, the beam of sunlight will fall at *D*; but for every degree that the sun rises above the horizon, the projected sunbeam will drop 2° below D along the inside of the ring.

If we were to cut the ring at its point of suspension, M, and straighten it out into a flat band, as in the lower part of Figure 16.7, the cut at M would be represented by the two ends of the flattened band at AB and CD. The entire length of the band (AC or BD) would represent the 360° of the circle, so each degree would take up $\frac{1}{360}$ of the length of the band. Since in the upper circle MD and ME are each taken equal to the latitude (here 42°) the point D of the upper circle will lie on a line EF of the band at the bottom, $\frac{42}{360}$ of the length of the band from AB—and this line EF will represent an altitude of zero. The aperture at E in the upper diagram will now appear at G in the lower part, also $\frac{42}{360}$ of the way from CD to AB.

Divide the band itself into twelve narrow strips, representing the months of the year. Then find for any hour of any day the sun's altitude, by methods described earlier in this chapter when we were considering the shepherd's dial. For example, on September 1 at 8 A.M. and at 4 P.M. in latitude 42° N the sun's altitude is 27.8°. But since the inscribed angle is measured by half the subtended arc, the 27.8° of solar altitude at BOE in the upper diagram will be measured by an arc of 55.6° at DF, and we should measure $\frac{55.6}{360}$ of the length of the band, starting at EF, along the lengthwise line for September 1. In this way we find where each hour line crosses each month line, and connect them with smooth curves for the hour lines. Two of these hour lines, those for 12-o'clock noon and for either 8 A.M. or 4 P.M., are shown drawn to scale in the lower part of Figure 16.7.

It will immediately become clear that these perforated ring dials, while easy to lay out in theory, will be too crowded for accurate reading unless the ring is very large, in which case they lose their portability.

The Sun Watch. For some years a small and handy folding pocket sundial has been available through the Boy Scouts of America. Known as the "sun watch," the instrument is housed in a thin metal case which, when opened, discloses a horizontal sundial with three sets of concentric hour lines for latitudes 35°, 40° and 45°. The gnomon is also adjustable and marked for the same three latitudes, which cover roughly 75% of the area of the United States including most of the heavily populated areas. The base of the instrument contains a tiny compass for aligning the dial, and the cover contains a list of over 40 major U.S. cities with their latitudes, longitudes, longitude corrections, and compass variations. There is also a table showing values of the equation of time for selected dates. The instrument is shown in Figure 16.8; and any boy lucky enough to own one has far more than an interesting toy. The large amount of useful information tabulated inside the cover adds greatly to its value.

The Capuchin Dial. Old books on sundials often described what they referred to as "the dial on a card"—sometimes called a "Capuchin dial" for reasons which will appear later. This is a reasonably accurate little timekeeper which any boy can make with materials readily available, although younger boys may need some help in lying out the angles.

To construct the dial we take the following steps, reference being made to Figure 16.9.

(1) On the small card on which the finished dial will appear, draw *GH* parallel to the top of the card.

(2) Draw *EK* perpendicular to *GH* intsersecting it at *P*. This will become the 6-o'clock line.

(3) With *P* as the center and with any convenient radius draw the semi-circle *GKH*.

(4) Beginning at *G*, divide this semi-circle into twelve equal arcs of 15° each at a, b, c, d, etc.

(5) Through these points of division on the arc, draw lines parallel to *EK* for the hour lines.

(6) Draw *GL*, making angle *PGL* equal to the latitude.

(7) At the point where *GL* intersects *EK* draw *VW* perpendicular to *GL*.

(8) Draw *GV* and *GW* making angles *LGV* and *LGW* equal $23\frac{1}{2}°$.

(9) Lay off along *VW* the lines marking the first days of the various months, so that they make angles with *GL* equal to the sun's declination on the first day of the month. For example, the little line between *J* and *F* on the scale marks the first day of February, and if this line were extended to *G*, the angle with *GL* would be 17°20′, which is the sun's declination on February 1. (See Appendix, Table A.2.) Negative solar declinations lie from *L* toward *V*; positive ones from *L* toward *W*.

(10) With *V* as the center and *VG* as the radius draw the arc *GQ* for the Tropic of Capricorn.

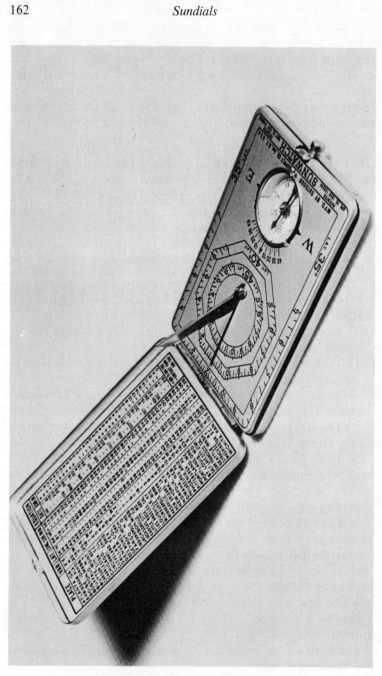

FIGURE 16.8 Portable Boy Scout sundial.

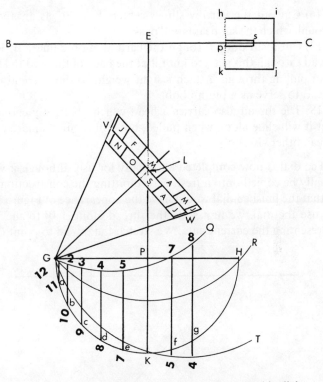

FIGURE 16.9 Diagram for construction of a Capuchin dial.

(11) With *L* as the center and *LG* as the radius draw the arc *GR* for the equator.

(12) With *W* as the center and *WG* as the radius draw the arc *GT* for the Tropic of Cancer.

(13) Draw the hour lines from arc *GQ* to arc *GT* and number them as shown in the figure. The morning hours are shown on the lower arc and the afternoon hours along the upper arc. A few of the hour numbers have been omitted in the diagram to avoid confusion.

(14) Draw the shadow line *BC*, near and parallel to the upper edge of the card.

(15) Lay out a small rectangle, *hijk*, centered on and toward the right end of *BC*. Cut quite through the card along the three sides of this rectangle, *hi*, *ij* and *jk*, and fold the rectangle out at right angles to the card along the dotted line *hk*.

(16) Cut out the narrow slit *ps* centered on *BC* as an aperture through which the sunbeams will pass.

(17) Cut a slit quite through the card along the line *VW*. Run a thread through this slit and knot it at the back of the card to keep it from pulling through. Attach a light weight to the free end of the thread to serve as a plumb bob.

(18) The thread also carries a bead which fits tightly enough so that it will slide along when pulled, but will retain its place on the thread otherwise.

The dial is now completed and ready for use, although it would usually be copied onto a fresh card omitting the construction lines so that the finished dial would have the appearance of Figure 16.10. To use the dial, we first slide the thread along *VW* to the point representing the current date. We stretch the thread to point *G* (the

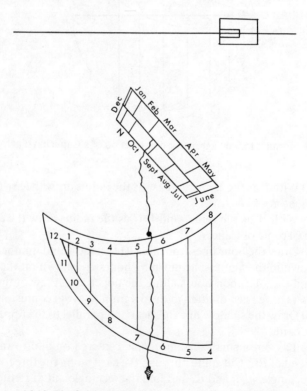

FIGURE 16.10 The finished Capuchin dial constructed from Figure 16.9.

12-o'clock point) and slide the bead along the thread to that point. With the gnomon (the rectangle *hijk*) folded out at right angles to the card we hold the card itself in a vertical plane pointed toward the sun and tilted until the shaft of sunlight falling through the slit at *ps* falls along the shadow line toward *E* and *B*. The thread, hanging free but pulled into a vertical position by the weight, will carry the bead among the hour lines, where it will mark the hour of the day. The bead will fall on arc *GQ* on December 21, on arc *GT* on June 21, and on arc *GR* at the times of the equinoxes.

One can also tell the time of sunrise and sunset from this dial. With the upper end of the thread set at the appropriate date, pull the thread taut and swing it until it is parallel to the hour lines. The thread will now cut the morning hours at the time of sunrise and the afternoon hours at the time of sunset for the particular date. The theoretical basis for this dial is ably explained in an article by Frederick A. Stebbins entitled "A Medieval Portable Sun-dial" in the April, 1961, issue of the *Journal of the Royal Astronomical Society of Canada*, and the dial is also discussed in the "Amateur Scientist Department" of the *Scientific American* for May, 1966.

The Capuchins were an order of Italian monks established in the sixteenth century, and the Capuchin dial took its name from a fancied resemblance of its curves to the hoods worn by these monks, as shown in Figure 16.11.

FIGURE 16.11 Capuchin dial inverted and with additions to show the origin of its name.

A Very Early Portable Dial. When the Danes ravished Canterbury, England in the year 1011, they imprisoned Archbishop Alphege and murdered him in prison the following year when he refused to pay a ransom. When the Garth Monastery at Canterbury was restored in 1938 his tomb was opened, and in it was found a little silver sundial hung from a golden chain. The gnomon was also of gold with one end fashioned into the head of an animal with tiny precious stones inset for the eyes. A reproduction of this early dial, now in the Hellmut-Kienzle Uhren-Museum at Schwenningen, West Germany, is shown in its actual size in Figure 16.12 by courtesy of my friend, Herr Paul Melchger. The original dial is obviously over nine and a half centuries old.

Each face of the dial is divided into three columns, and each column carries the names of two months. The gnomon is a removable pin which is ordinarily stored in a hole in the base; but when the dial is in use the gnomon is removed and thrust through the hole at the top of the column for the appropriate month. The dial, hanging by its chain, is then turned toward the sun until the shadow of the gnomon falls on the column below. Two dots appear in each column. When the shadow falls on the lower of these dots it is noontime, while at

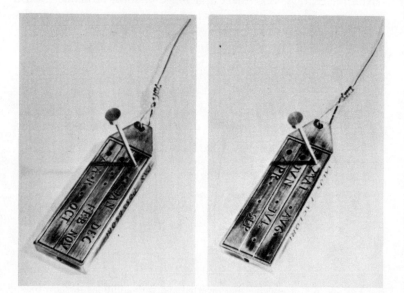

FIGURE 16.12 The two sides of an ancient portable sundial.

midmorning and midafternoon the shadow falls on the upper dot. Thus the dial showed the times of the medieval unequal canonical hours which were set aside for prayers. We can imagine the Archbishop holding his little dial before him as he called his monks to their devotions. Inscriptions on the sides of the dial invoke blessings on the unknown maker, and carry wishes for peace to the possessor.

Time-Telling at Night. It may seem peculiar to include in a book on sundials a brief reference to instruments designed to tell time at night—but the apparent motions of the moon and stars bear some resemblance to the motions of the sun, and they, too, have been used for timekeeping.

Time by Moonlight. For a week or so just before and after the time of full moon, the moonlight is bright enough to cast a shadow on a sundial, and this moon shadow can be used to get a rough estimate of the time of night if we make proper corrections. The motions of the moon are highly complicated, but for our purposes we shall be close enough if we say that the moon "runs slow" by about 2 minutes an hour or 48 minutes a day. A week before full moon the time shown on the sundial by moonlight is about $5\frac{1}{2}$ hours "fast"; on the night of full moon the time shown by moonlight is correct; and by a week after full moon the "moon time" is about $5\frac{1}{2}$ hours "slow." The amounts of the correction during the weeks before and after full moon are as follows:

days from full moon:	0	1	2	3	4	5	6	7
amount of correction:	0:00	0:48	1:36	2:24	3:12	4:00	4:48	5:36

We must remember that on days before full moon the time shown by moonlight is "fast," and on days after full moon it is "slow." Suppose that the shadow cast by moonlight on a sundial indicates that the time is 10:30, and that it is 2 days after full moon, we apply the correction of 1 hour and 36 minutes to estimate that the time is really about midnight. Strict application of our rule would give the time as 12:06 A.M., but our rule is too rough to warrant such misleading precision. Had our observation been made 2 days before full moon we would have estimated that the time was really about

9 P.M., since strict application of the rule would have given a time of 8:56 P.M. Some sundials have actually carried tables of these corrections for use by those few observers with nocturnal habits who might understand what the table is all about—and any dyed-in-the-wool sundial enthusiast may take pleasure in checking his dial by moonlight on occasion merely as an experiment. But as a practical timekeeper this type of "moon dial" leaves much to be desired, both because the corrections are at best very rough, and because there are relatively few times at night when the sky is clear enough and the moon bright enough for us to get a readable shadow on the sundial.

A similar argument is sometimes made about the sundial itself. Even during the daytime hours it is not always sunny, and there is not always a shadow on the sundial. The amount of sunshine which we receive depends in part on our latitude and in part on local weather conditions. Figure 16.13 shows two maps prepared by the U.S. Weather Bureau showing the average numbers of hours of sunshine actually received during the summer and the winter months in various parts of the continental United States. For example, in Chicago we can expect an average of between 4 and $4\frac{1}{2}$ sunny hours on winter days, and 10 to $10\frac{1}{2}$ hours on summer days. Since sundials today get their principal use in summer months when people are out of doors, the maps should give some encouragement to sundial users, since they indicate that most parts of the country can expect sunshine for a large proportion of those summer hours when the sun is above the horizon.

The Nocturnal. The tossing deck of a sailing vessel was ill adapted for the use of a sundial, and sailors needed some other means of keeping track of the time. They early developed reasonably satisfactory methods for use in clear daytime weather, based on observations of the sun's altitude; and for several centuries they depended at night on observations of the stars. They noted that the northern constellations appeared to rotate around the Pole Star approximately once each day, like a huge hand on a celestial 24-hour clock; and mariners memorized the positions of the more conspicuous northern stars at various times of night at the various seasons of the year. And centuries ago they developed a simple small instrument to aid them in making their observations—an instrument to which they gave the Latin name *horologium noctis*, but which Englishmen called the "nocturnal." This instrument is so easy to make that

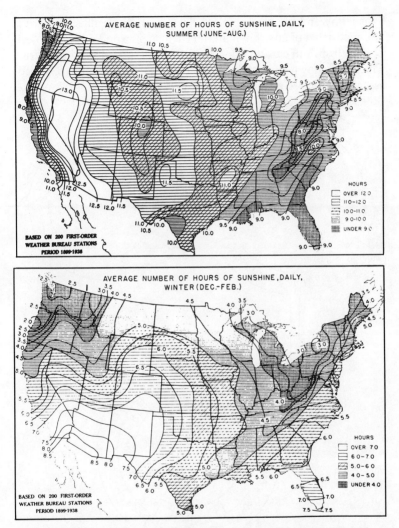

FIGURE 16.13 Amounts of summer and winter sunshine in the United States.

many a Boy Scout has amused himself on an overnight hike by reading the time from the stars by means of his own copy of this medieval scientific instrument.

While most of the ancient nocturnals which have survived are beautifully engraved on copper or brass, and are anxiously sought by wealthy and knowledgeable collectors and museums, they may be constructed of any thin material such as plywood or prest-board or even from stiff cardboard. The instrument is made of three parts, which are shown separately in the upper part of Figure 16.14 and assembled for use at the bottom of that figure.

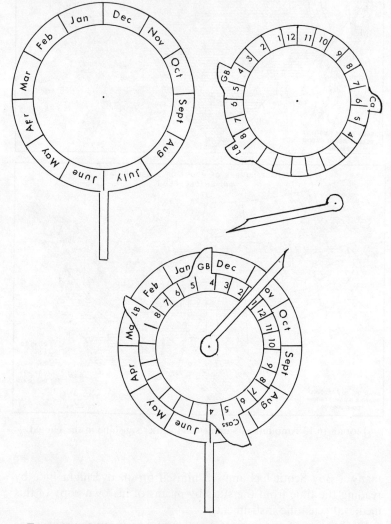

FIGURE 16.14 The nocturnal (bottom) and its three constituent parts (top).

We first cut out two circular plates of convenient size, the larger perhaps 3 inches and the smaller about $2\frac{1}{2}$ inches in diameter. The larger disk has a handle at one edge which is either cut out as an integral part of the disk or attached at the back subsequently. This larger disk is divided around its edge into twelve divisions, starting with a division representing the first day of January directly opposite the handle. The twelve divisions represent the twelve months and run counterclockwise around the edge of the disk. There will be little error if they are laid out of equal size, each subtending an arc of 30°; but if one wishes to be precise it is not difficult to proportion them to the varying lengths of the months. Counting 365 days to the year, we would assign to January $\frac{31}{365}$ of the 360° of the circumference or 30.6°; to February we would assign $\frac{28}{365}$ or 27.6°, and so forth. Figure 16.14 is drawn to such a small scale that only monthly divi-

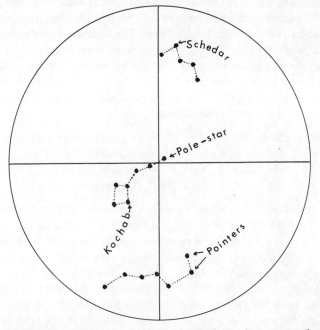

FIGURE 16.15 Some of the stars of the north sky as they appear about 9 P.M. in mid-November. The upper five stars constitute part of the constellation Cassiopeia; the middle group part of Ursa Minor (the "Little Dipper" or "Lesser Bear"); and the seven stars at the bottom are part of Ursa Major (the "Greater Bear" or "Big Dipper" or the "Plow").

sions are shown, but in practice one would insert subdivisions for every 5 or 10 days.

We next cut out the smaller disk of such a size that the monthly divisions which we made around the edge of the larger disk will be plainly visible when the smaller one is centered on top. If, for example, the diameter of the smaller disk is $\frac{1}{2}$ inch smaller than that of the larger disk, there will be a $\frac{1}{4}$-inch margin of the larger disk showing all the way around the smaller disk when they are assembled. The second or smaller disk has one or more "teeth" on its outer edge which must be located accurately and which should reach out at least to the outer edge of the larger disk and perhaps a little farther, as shown in the lower part of our figure. Before these teeth can be accurately located we divide the circumference of the smaller disk into 24 equal divisions, each of 15°, representing the 24 hours of the day. One of these divisions is selected as the point for 12 o'clock midnight, and the other hours are then numbered in sequence running counterclockwise. The nocturnal was sometimes made with but one "tooth," and such an instrument can be made today if one wishes; but since it will soon become apparent that the instrument is more useful with added teeth, and since there is little difficulty in adding them, we shall describe an instrument with three teeth which are located as follows:

(1) A tooth marked "GB" 63.4° counterclockwise from the midnight division at a position corresponding to 4:14 A.M.

(2) A tooth marked "LB" 124.4° counterclockwise at 8:18 A.M.

(3) A tooth marked "Cass" 94.7° clockwise at 5:41 P.M.

The third part of the nocturnal is a cursor or ruler, pivoted at one end. One of its edges (called the "fiducial edge") is so laid out that if it were prolonged it would pass through the center of the pivot.

The instrument is assembled with the larger disk at the bottom, the smaller disk in the middle, and the cursor on top, as shown in the lower part of Figure 16.14. The three parts are held together by a hollow rivet or (perhaps easier to come by) a short nipple of $\frac{1}{2}$-inch galvanized pipe with a nut screwed on each end to hold the disks in place. Any electrical supply house can furnish nuts of the proper size and thread to fit $\frac{1}{2}$-inch pipe. The assembly must be loose enough so that the disks and the cursor may be turned upon each other, but snug enough so that they will hold their position once they are properly aligned.

In use the instrument is first set for the date for which it is to be used and for the constellation which is to be observed. If we decide to observe the position of the Big Dipper or Great Bear, we set the tooth labeled "GB" on the smaller disk at the appropriate date.[2] We then hold the instrument at arm's length with the handle pointing vertically downward and with the plane of the disks perpendicular to our line of sight to the North Star. We sight the North Star through the hole in the central pivot and, keeping the star centered in the hole, swing the cursor until its fiducial edge touches the appropriate part of the constellation which we are observing. The fiducial edge will then cut the divisions on the smaller inner disk at the approximate local mean time. Thus the instrument at the bottom of Figure 16.14 is set as it would have been if we had made an observation at about 1:25 A.M.—and if our observation had been on the "Pointers" it was made on about December 29; if on Kochab about March 1; and if on Shedar about July 22.

---

[2] The labels on the three teeth of the smaller disk have the following meanings: "GB" means "Great Bear" or Ursa major, and is used for observations on the "pointers" of the "Big Dipper"; "LB" means "Lesser Bear" or Ursa minor, and is used for observations of the star Kochab which lies at the extreme end of the bottom of the bowl of the "Little Dipper"; and "Cass" means "Cassiopeia," and is used for observations on the star Shedar which lies at the right of the base of the distorted "W" in the constellation Cassiopeia. These stars identified in Figure 16.15 which shows the northern sky as it would appear on an autumn evening. The advantage of having more than one tooth on our nocturnal is apparent when we realize that if one of the constellations is below the horizon or obstructed by trees or a passing bank of clouds, we can observe one of the other constellations by setting the appropriate tooth opposite our date.

17
The Armillary Sphere

Ancient astronomers used celestial globes which depicted the heavens much as modern "globes of the world" show the seas and the continents. The celestial globe showed the constellations and their brightest stars, and the imaginary circles of the heavens corresponding to the meridians and parallels on terrestrial globes. For many purposes the body of the globe was omitted and a mere skeleton used instead, consisting of an assembly of rings representing the principal circles of the heavens. Such an assembly came to be called an "armillary sphere" from the Latin word *armilla*: a bracelet or ring. Some armillary spheres were highly complicated with dozens of rings; but for most purposes they were reduced to three or four rings, and in this form they became one of the most handsome and attractive sundials, like the one which stands in front of Old Main Building at Pennsylvania State University, shown in Figure 17.1. In its simplest form the armillary sphere is but an extension of the equatorial dial described in Chapter 4, with two rings representing the ecliptic and the meridian, and with a rod passing through their common center representing the earth's axis. A third ring is usually added representing the horizon, and our description will cover such a three-ring sphere.

Methods of Measurement. The rings or circles comprising an armillary sphere are most naturally measured in circular units—in degrees, minutes, and seconds of arc. But even if our original computations are in such units, we will often find it simpler to apply them in the more familiar units of linear measure—feet and inches, or centimeters. In any particular case conversion from one of these systems to the other is easy. Imagine a ring 11 inches in diameter, cut and straightened out into a band. This band will be just over 34.5 inches long (11 times pi). This length of 34.5 inches represents the

FIGURE 17.1 Armillary sphere, Pennsylvania State University.

entire 360° of the original circle, so each degree is equivalent to $\frac{345}{360} = 0.096$ inches. For this ring we can convert degrees to inches by multiplying by 0.096, or inches to degrees by dividing by 0.096 (or by multiplying by its reciprocal, 10.42). When in our later directions we say to lay off a distance of, say $23\frac{1}{2}°$, we can, if we prefer, measure a distance of 2.256 inches along the arc of the ring.

Laying Out the Rings. For a metal worker, the armillary sphere is a simple and rewarding project. Armillary spheres may be constructed so that they are universal (usable in all latitudes), but they are usually made for some specific latitude. This is especially true if they include a horizon ring. While commercial designers like Kenneth Lynch of Canterbury, Connecticut, make splendid ornamental spheres 15–20 feet or more in diameter, and while the amateur artisan may make a sphere 3 or 4 feet in diameter by welding together the metal tires of old wagon wheels, we shall illustrate the layout of a sphere with rings 1 foot in diameter as being more nearly the size of a garden sphere. The rules can be adjusted easily for rings of other sizes.

We start with the band which will bear the hour numbers and represent the equator. It will be the innermost of the rings which make up the sphere, and since it is 12 inches in diameter we shall need a band of metal 37.7 inches long. For good proportions it might be roughly $\frac{3}{4}$ inch to 1 inch wide and $\frac{1}{16}$ inch thick—although the width and thickness will depend on the stock of metal at hand. Since the entire circle will measure 37.7 inches, each degree will correspond with a linear measurement of 0.1047 inches. The hour marks should be 15° apart, or 1.57 inches. We place the hour line for 12 noon at the middle of our band of metal, and lay out the other hour lines toward the left and right 1.57 inches asunder, with the morning hours at the left and the afternoon hours at the right as shown in Figure 17.2. Since the sphere will show the time whenever the sun is above the horizon, we should include hour lines from the time of earliest sunrise to the time of latest sunset for our latitude. The hour lines of 6 A.M. and 6 P.M. should be just 18.85 inches apart—

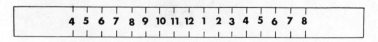

FIGURE 17.2 Equatorial band for armillary sphere.

just half the length of the band, so that when the band is bent into a circle with the hour marks on its inner surface these two hour lines will lie just opposite each other with the 12-o'clock hour line at the bottom just half way between them. The ends of the band are now welded together, or otherwise connected, to make a circle 12 inches in diameter.

Our second ring is the horizon ring, which should fit snugly over the equatorial ring and must, therefore, be a very little bit longer. The amount of the added length will depend on the thickness of the metal stock being used, and one should try the strip of metal around the outside of the equatorial ring before cutting it to length so that the two rings will nest snugly together. Since there is no marking on the horizon ring the band can be welded into a circle as soon as it has been cut to the proper length. Finally we make the meridian ring, which is cut to the proper length to fit snugly around the horizon ring, being the longest of the three rings.

The three rings must now be combined in their proper positions. The meridian ring is to be attached to a pedestal, and the attachment is usually made at the point where the ends of the band were welded together, as at point A in Figure 17.3, which shows an armillary sphere made by a Connecticut amateur craftsman in his home workshop. In this figure, the circle $ABCDEFG$ is the meridian, $DHGJ$ the equator, and $CHFJ$ the horizon.

We now attach the horizon ring at points C and F. Each of these points should be 90° from A—or 9.423 inches. And the equatorial ring is then attached to the meridian ring at points D and G. In any armillary sphere point G should be distant from A by the latitude; and point D should be distant by 90° plus the colatitude. If, for example, our sphere is made for latitude 42°, AG will be 42° or 4.40 inches and AD will be 138° or 14.45 inches. We next connect the horizon and the equator at points H and J, halfway between D and G and also halfway between C and F. In other words, $DH = HG = CH = HF = 90° = 9.42$ inches. Finally we attach the gnomon at points E and B, with AB equal to the colatitude and AE equal to 90° plus the latitude. For our case, $AB = 48° = 5.03$ inches and $AE = 132° = 13.82$ inches. We can check our measurements around the meridian ring, if we wish, by the fact that on any armillary sphere:

$$AG = FE = BC = \text{latitude}$$
$$GF = AB = CD = \text{colatitude}$$

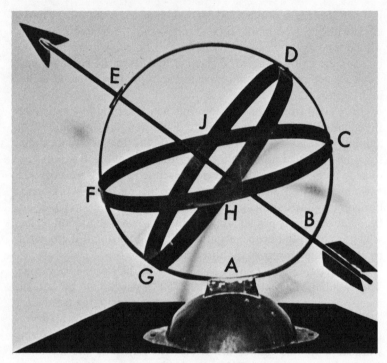

FIGURE 17.3 Homemade armillary sphere.

The gnomon is usually a long, thin rod which runs through holes pierced in the meridian ring. The shadow of the gnomon, cast among the hour lines on the equatorial ring, shows the time. Just at noon the shadow of the meridian ring itself will fall across the gnomon and also across the 12-o'clock line. At the times of the equinox, the sun will be on the celestial equator, and hence in the plane of the equatorial ring of the armillary sphere. The shadow of the upper part of the ring (*D* in Figure 17.3) will fall on the lower part of the ring, and the hour lines will be shaded. If one omits the horizon ring (as he might do if he were welding together two large iron tires from wagon wheels) he will have distances $AG = \phi$; $GE = 90°$; $AB = \text{co-}\phi$; and $BD = 90°$.

The Sun's Declination. At the times of the equinoxes the shadow of the equatorial band falls at the center of the gnomon, half way between *E* and *B* in Figure 17.3. In the summer, when the sun is high, the shadow falls farther down the gnomon closer to *B*, while in winter it falls closer to *E*. One can easily calibrate the gnomon rod to show the sun's declination by the position of the shadow on the gnomon.

Starting at the center of the gnomon, just half way between *E* and *B*, we measure up toward *E* a distance equal to half the width of the equatorial band. This is the zero point for our calibration, from which we measure upward toward *E* for southerly and downward toward *B* for northerly declinations of the sun. Given the radius, *R*, of our equatorial circle, the distance, *D*, which we measure from the zero point for any particular solar declination, *d*, is

$$D = R \tan d.$$

In our illustration, where the radius of the equatorial circle is 6 inches, if the sun's declination is 10° north (as it is about April 16 and August 28 each year) the equation tells us that

$$D = (6)(\tan 10°) = (6)(0.176) = 1.056,$$

so we would measure 1.056 inches from our zero point toward *B* and mark the point where the shadow would fall at these two times of year. Similarly we would mark another point 1.056 inches above the zero point for a southerly declination of 10° (February 23 and October 20 each year). The scale will usually be marked for each 5° or 10° of solar declination, and we must read the declination from the upper edge of the shadow of the equatorial band, since we chose as the zero mark for our scale the point on the gnomon corresponding to this upper edge.

The home craftsman who works in metal will have little difficulty in making an armillary sphere which will be a joy to him and an ornament to his garden. Figure 17.4 shows such a sphere, more ambitious, to be sure, than would ordinarily be made for a home garden, which was designed and built for the Old Mystic Seaport by Edwin Pugsley; but a modest sphere like that of Figure 17.3 is a project well within the reach of a high school manual training student or of the craftsman who works with metal projects evenings in his garage.

FIGURE 17.4 Armillary sphere at Old Mystic Seaport.

18
Memorial Dials

Any well-designed sundial makes a fitting memorial, and all across the country, on college campuses and village greens, we see sundials dedicated to the memory of distinguished alumni or faculty members, or to founding fathers and worthy citizens. A sundial makes a much more imaginative memorial to veterans who gave their lives in battle than does the standard bronze plaque—and the plaque, if desired, can be set into the pedestal on which the dial stands. Many an English village has had from the middle ages a "town post" with dials on the four sides, antedating the development of the town clock. But there are two or three sundial designs which, by their nature, are especially suited for use in a cemetery, and we shall treat of such dials in this chapter.

Headstone with a Sloping Surface. Figure 18.1 shows a headstone which has been cut to leave a sloping surface at the center of the top. The edges of this surface are gnomons which cast shadows among hour lines which are cut into the level spaces at the ends of the top. The dials are separated halves of an ordinary horizontal dial, so the upper slope must be equal to the latitude of the place. Any ordinary stonemason can do the necessary work, but will need explicit detailed instructions to guide him. The directions are simplified if the headstone faces due south; but if, as will often be the case, the stone faces at an angle, the sloping tablet must also stand at an angle to the main body of the headstone in order that its edges, which serve as gnomons, may lie in the meridian. Figure 18.2 shows a headstone which faces somewhat to the east of south.

The Cross Dial. In Chapter 16 we referred to a portable dial in the shape of a cross; but since cruciform dials are especially suited as cemetery memorials we postponed the details to this point. The

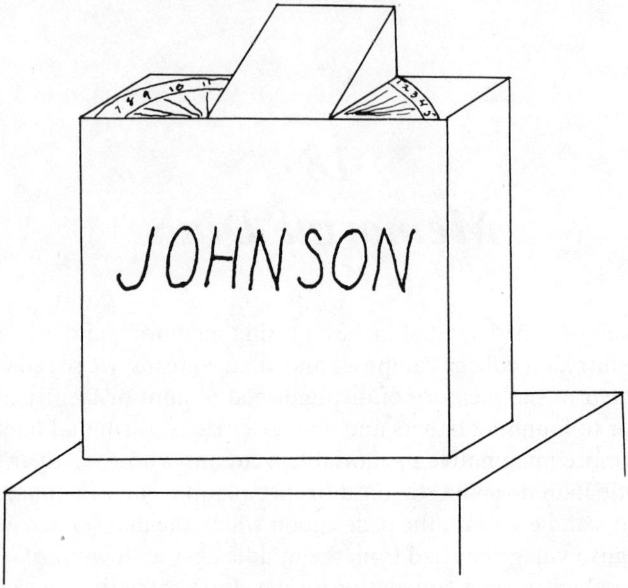

FIGURE 18.1 Cemetery headstone with sundial. Stone faces due south.

general appearance is shown in Figure 18.3. The upper part of the gravestone is cut in the shape of a cross which faces toward the north, but with the top tipped back toward the south until the cross lies in the plane of the Equator. Thus we are looking at the north side of the stone in the figure.

The cross can be considered as composed of a mainshaft to which three cubes have been attached—two at the sides to form the cross arms and the third at the top. As the sun swings across the sky in the course of the day, the shadows of these cubes will fall upon the main shaft or upon each other. Thus just at noon, when the sun is due south, the beams will fall directly on top of the topmost cube, and it will cast no shadow. But as the sun passes toward the west (the right in our figure) the shadow of this topmost cube will appear on the top of the eastern (lefthand) cube, and will move across it for the next two or three hours.

Since the cross itself lies in the plane of the Equator, the edges of the cubes, which will serve as gnomons, lie parallel to the earth's axis. Thus we can make ordinary polar dials for the upper faces of the two side cubes, and direct vertical east or west dials for the sides

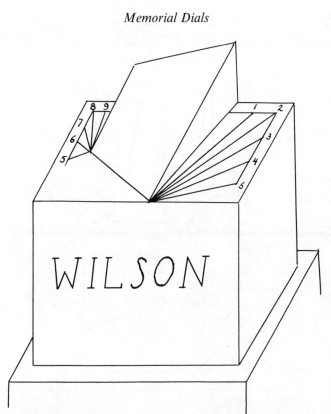

FIGURE 18.2 Cemetery headstone with sundial, stone facing southeast.

of the main shaft and the sides of the upper cube. Since the cubes all have the same dimensions, the spacing of the hour lines will be repeated on each cube, and the arrangement of hour lines will be as shown in Figure 18.4. At 6 A.M. the sun shines perpendicularly on the eastern cube, at 6 P.M. on the western cube and at noon on the top of the upper cube. In computing the hour angles for the shadow cast by any one of these cubes, we take the sun's hour angle from the time when the sun shone perpendicularly on that cube. Thus the edges of the eastern cube will cast shadows on the topmost cube from sunrise until 6 A.M. and then down the eastern side of the main shaft of the cross from 6 A.M. until late in the morning. (See Figure 18.4.) For these hour lines, with the shadow cast by the eastern cube, we use the sun's hour angles from 6 A.M. Thus 5 A.M. and 7 A.M. would be hour angles of one hour or 15° as far as this cube is concerned—but when we are

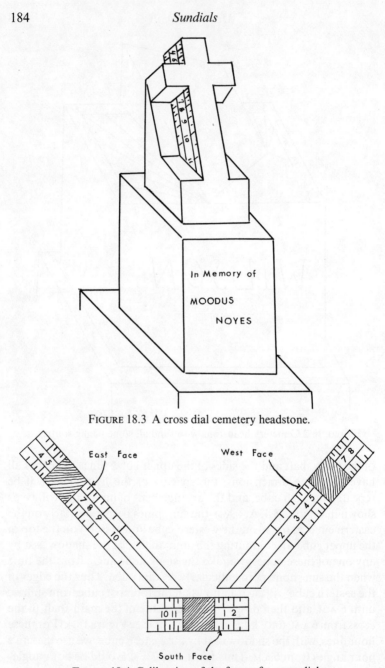

FIGURE 18.3 A cross dial cemetery headstone.

FIGURE 18.4 Calibration of the faces of a cross dial.

considering the shadow cast by the upper cube, where the sun is perpendicular at noon, the hour angle of one hour or 15° would represent either 11 A.M. or 1 P.M.

The length of the shadow at any hour angle will vary according to the tangent of that hour angle, as shown in Table 18.1. The values of the sun's hour angle are given at intervals of a quarter hour, and the lengths of the shadow are in units of the side of the cube—that is, we multiply each length given in Table 18.1 by the length of the edge of the cube to convert them to inches or to centimeters or to such other units as we used in measuring the edges of the cube. For example, suppose the sides of the cubes are 5-inch squares, and that we want the hour line for 8 A.M. The shadow will be cast on the main shaft of the cross by the eastern cube (see Figure 18.4) and the time is 2 hours or 30° from the time when the sun shone perpendicularly on this cube. Hence the distance of this hour line from the base of the cube will be (0.577)(5) or 2.885 inches. Comparable procedures are used to locate the other hour lines. Reference to the figure should give such further explanation as may be necessary.

TABLE 18.1
LENGTHS OF SHADOWS ON A CROSS DIAL AT VARIOUS HOUR ANGLES, IN UNITS
OF THE LENGTH OF THE EDGE OF THE CUBES

sun's hour angle	length of shadow	sun's hour angle	length of shadow
0.00°	0.000	30.00°	0.577
3.75°	0.065	32.75°	0.668
7.50°	0.132	37.50°	0.767
11.25°	0.199	41.25°	0.877
15.00°	0.268	45.00°	1.000
18.75°	0.339	48.75°	1.140
22.50°	0.414	52.50°	1.303
26.25°	0.493	56.25°	1.497
		60.00°	1.732

Star of David Dials. The six-pointed Star of David can be adapted for use as a memorial sundial comparable to the cross dial just described. The star itself consists of six equilateral triangles set on the sides of a regular hexagon. The "points" of the star are 60° angles, and the obtuse angles formed where any two points join are angles of 120°. The star lies in a sloping position facing the north, with the plane of its surface parallel to the equator. This means that this surface is tipped out of the vertical by an angle equal to the latitude. Figure 18.5 shows the completed monument, with the topmost point of the star pointing upward toward the south, and with east at the left and west at the right.

As the sun moves across the sky the shadow of the various points of the star are cast onto the sides of adjacent points, and the proportions are such that it takes just two hours for the shadow to traverse the side of each point. As the shadow leaves the hour scale at any one point, another shadow starts its journey on the hour scale of some other point, so that whenever the sun is shining there is a shadow on some part of the star to denote the time.

The underlying theory will be apparent from Figure 18.6. The upper part of this figure shows the star as it would appear if we looked perpendicularly at its surface. The sides of the six points have been designated with the letters from *a* through *l* so that we may identify them. The lower part of the diagram shows these twelve faces, identified by the same letters, as they would appear if they could be peeled off the points of the star and spread out into a band—or, in this case, into two bands. The small letters in the lower

FIGURE 18.5 Star of David monument as viewed from the northwest.

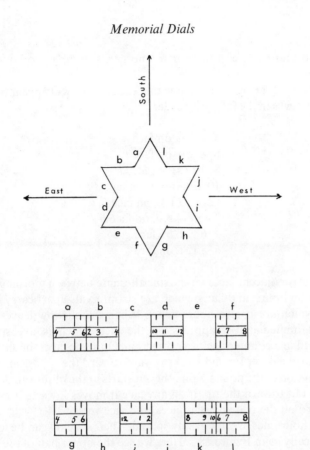

FIGURE 18.6 Star of David (top), and hour lines as they appear on the various faces (bottom). Hour lines are closely bunched at the obtuse angles of the star, but more widely spaced near the outer points.

part of the diagram identify the same sides of star points which they represent in the upper part. For an hour or so before noon, when the sun was just to the left of being "straight up" on the diagram, the extreme outer edge of face *b* was casting a shadow on face *d*; and immediately after noon, as the sun moves toward the right, the shadow of the extreme outer tip of face *k* starts to fall upon face *i*. In the "Land of the Midnight Sun," the summer sun would reach all twelve faces of the star at one time or another during the 24 hours; but in temperate latitudes some of the faces, toward the northern side of the star, will never catch the sun's rays. Consequently these faces appear blank in the bottom of Figure 18.6. The shadow will move from one face of the star to another, spending two hours on each face.

A study of the figure will show that the shadow will appear on the various faces in the following order:

> 4–6 A.M. on face *a*
> 6–8 A.M. on face *f*
> 8–10 A.M. on face *k*
> 10 A.M.–noon on face *d*
> noon–2 P.M. on face *i*
> 2 P.M.–4 P.M. on face *b*
> 4–6 P.M. on face *g*
> 6–8 P.M. on face *l*

Obviously adjacent faces on the star alternate between morning and afternoon hours, and faces which are six intervals apart carry hour lines which are twelve hours apart. When the shadow leaves any face it immediately reappears on the face which is five spaces removed in a counterclockwise direction. On some faces the shadow starts at the outer tip and works inward toward the obtuse angle at the junction of the points—on others it starts at the obtuse angle and moves out toward the tip. If on any face it moves inward, it moves outward on the two adjacent faces, and vice versa.

The hour lines are not evenly spaced. Their spacing can be found graphically as in Figure 18.7. Here we see the upper half of the star, with the topmost point labeled *cOg*. If this point is formed by faces *a* and *l* of the preceding figure, then the line *eOS* lies in the meridian with *S* at the south. At noon the sun shines in the direction *S* to *e* and illuminates the entire faces *ac*, *cO*, *Og*, and *gi*. Let us draw arc *AcgB* centered at *O* and lay off upon it 15° arcs to show the sun's position at each hour. At 8 A.M. it will shine in the direction *Oi*, and then at successive hours its direction will change to *Oh*, *Og*, *Of*, *Oe*, *Od*, etc. It will move across the face *gi* of Figure 18.7 during the hours from 8 A.M. to 10 A.M., and if we continue the lines beyond the arc to the points where they strike the star at *g*, *h* and *i* we shall have the positions of the hour lines for 10 A.M., 9 A.M. and 8 A.M. Similarly we can find the positions of the other hour lines as they are shown in Figure 18.6. If you prefer, you can calculate the distances of the hour lines, using the tangents of the angles as in Chapters 8 and 9. You will find that if you start at the inner obtuse angle on any face and work

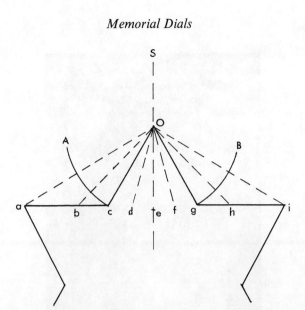

FIGURE 18.7 Finding the hour lines on a Star of David dial.

out toward the pointed tip, calling the length of the face 1.000, you will reach the various quarter-hour marks at these distances:

to first quarter hour	0.079
to second quarter hour	0.165
to third quarter hour	0.260
to fourth quarter hour	0.366
to fifth quarter hour	0.488
to sixth quarter hour	0.629
to seventh quarter hour	0.797
to eighth quarter hour	1.000

Each of these figures must be multiplied by the length of the face (distance from *a* to *c* in Figure 18.7) in inches or centimeters to give the distances of the hour lines in inches or in centimeters.

The sloping face of the star makes an angle with the vertical equal to the latitude. All Star of David dials are alike except for their scale. A star dial which has been correctly calibrated can be used in any latitude if the face of the star is tipped to the proper angle and if the star is properly oriented with respect to the meridian. Figure 18.8 shows a Star of David gravestone in a cemetery in Rochelle Park, New Jersey, made after the author's design.

FIGURE 18.8 Star of David memorial dial designed by the author.

19
Practical Hints on Dialling

Most of this book has been concerned with explaining how to lay out the design for a sundial. Our original design is often cluttered with construction lines and for our final dial we trace off carefully a full-sized diagram like that of Figure 19.1, which is based on the horizontal dial of Figure 5.5 on page 42. If the gnomon is to have

FIGURE 19.1 Finished design of the horizontal dial derived from Figure 5.5 on page 42.

significant thickness, we split the diagram into two parallel parts, as in Figure 5.4 on page 41. The final diagram may well carry other lines, even up to the complication of Figure 15.1 on page 123, but this diagram should carry whatever lines and symbols are to appear on our final work.

The problem remains of transferring the design from paper into some permanent form in which it can be placed outdoors in the weather. Some portable dials, which could be kept indoors except during moments of observation, were made with the paper pasted or glued to a wooden base, in which case the problem was simple.

Sundials are made of a wide variety of materials, so artists and artisans with widely differing skills and backgrounds may each find a field challenging his abilities, with materials and methods which are familiar to him. The reader is urged to begin with the materials closest at hand, and those to which he is most accustomed.

Old dials were often etched on metal. Today the etching would usually be done on heavy-gauge brass or aluminum plate, and the worker who has mastered the technique may add baked-on enamel for the lines of his design. The metal worker with skill in casting may wish to make a dial of cast metal similar to those found in the stores, but fitted to his own latitude and embodying his own creative design. If he is not equipped to carry out the final steps he may content himself with making the pattern, which is then taken to a professional for casting.

Some craftsmen will prefer to work in ceramics. The making of a ceramic sundial does not require a potter's wheel, but uses the "slab" method. The design is either incised into the slab or applied to it with glazes, and the gnomon is applied much as one attaches a handle to a ceramic teacup. One should remember that most ceramics will not stand freezing and thawing, and should be brought indoors for the winter.

Old dials were often made of wood and attached to the outer walls of buildings. Wood is still a useful material for the purpose. Figure 19.2 shows a modern dial of wood which has withstood a number of New England winters and which should last indefinitely with an occasional coat of paint. Even the beginning craftsman can manage a wooden dial, and many a boy has found it an interesting and informative project. The overall size of the dial will depend on its location, but in any case one starts by making his wooden dial plate and giving it one or two coats of flat white paint. Paints with a

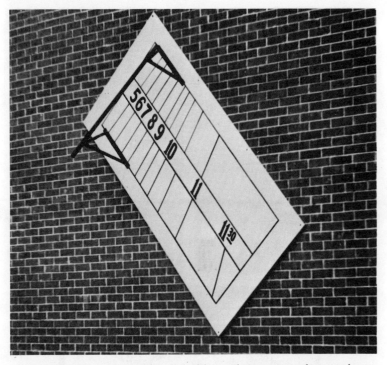

FIGURE 19.2 Large wooden dial with metal gnomon on the easterly
wall of a public building.

glossy finish should be avoided, since they do not take ink well. One
may wish to use what the paint stores call "white undercoater."
When this white undercoat is completely dry, the design is transferred to it, either with a ruling pen and black india ink (as in the
upper left of Figure 19.3) or with a modern indelible marking pen.
It is at this stage that provision must be made, if at all, to allow for
the width of the gnomon. If the dial is a small one, the gnomon may
be made of any metal stiff enough to resist bending. The small dial
of Figures 19.3 and 19.4 used a gnomon cut with a hacksaw from
one of the small steel plates used by electricians to cover junction
boxes. A small tab of metal (shown crosshatched in Figure 19.4) was
left projecting below the base line, AC, and a slot was cut through the
dial plate (upper right, Figure 19.3) into which this tab was inserted.
This slot is carefully placed so that the point of the gnomon (shown at
A in Figure 19.4) will rest exactly at the dial center from which the

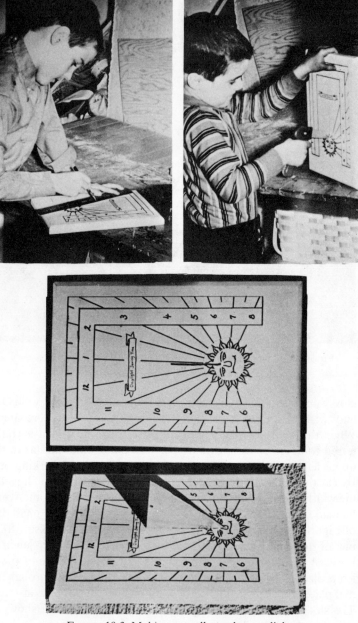

FIGURE 19.3 Making a small wooden sundial.

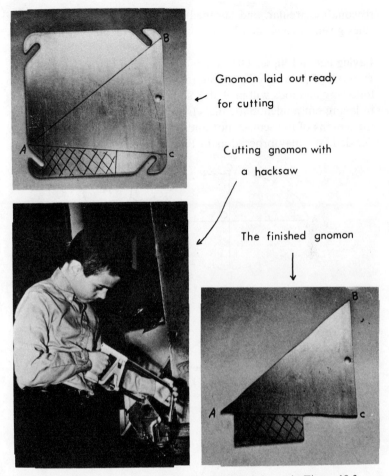

Gnomon laid out ready
for cutting

Cutting gnomon with
a hacksaw

The finished gnomon

FIGURE 19.4 Making the gnomon for the dial shown in Figure 19.3.

hour lines radiate. Since this is a horizontal dial, angle BAC in the figure must equal the latitude. Had the gnomon been cut out of wood it would have been screwed to the dial plate from below, or, better, one might cut a tenon on the base of the gnomon and insert it in a mortise cut through the dial plate. When all else is finished, the dial plate is given several coats of colorless outdoor spar varnish to protect it from the weather.

An armillary sphere can be made of wooden hoops from a barrel or cask, or from metal tires from wagon wheels. The hoops must be reasonably circular, and fastened together in accordance with the rules given in Chapter 17.

Laying out an Ellipse. Draftsmen are familiar with a number of methods of drawing ellipses, but for sundial purposes one of the two following methods will probably suffice. Ellipses are used primarily in laying out analemmatic dials (see Chapter 13) and in such cases the lengths of the semi-major and the semi-minor axes are already decided. We may proceed as in Figure 19.5, where the major axis,

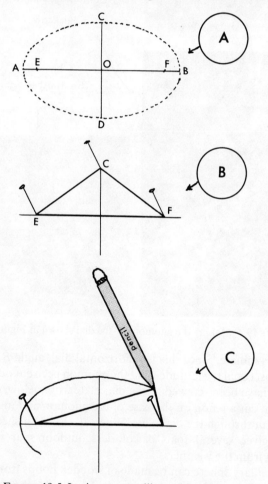

FIGURE 19.5 Laying out an ellipse with a loop of string.

AB, and the minor axis, *CD*, intersect at *O*. Setting one foot of the compasses in *C*, and with a radius equal to *AO*, draw short arcs intersecting *AB* at *E* and at *F*. These two points are the foci of the ellipse. Insert three pins in points *E*, *C* and *F*, as in part B of the figure, and encircle them with a loop of thread or stout string. Withdraw the pin at *C*, and put the point of a pencil inside the loop of string. Draw the loop tight with the pencil point, and, keeping the loop tight, swing the pencil point around the paper to give the ellipse seen under construction in part C of the figure.

Alternatively one can draw the ellipse by means of a trammell. In Figure 19.6 we have the major axis, *AB*, and the minor axis, *CD*, intersecting at *O*. We take a strip of cardboard, or a strip of wood like a yardstick, and mark on it points *E*, *F* and *G* so that *EG* is equal to the semi-major axis, *AO*; and so that *EF* is equal to the semi-minor axis, *OD*. We place the trammell on the two axes which we have already drawn of proper length (depending on the latitude), and we slide the trammell over the paper, keeping point *F* always on the major axis and point *G* always on the minor axis. Point *E* will then trace out the·desired ellipse. In laying out a large ellipse for a garden the author has made a makeshift trammell of a garden hose, with pieces of electricians' tape marking points *E*, *F* and *G*, and with two grandsons walking around the garden holding the hose taut while points *F* and *G* are moved along the axes already staked out on the ground. For a small ellipse points *E*, *F* and *G* may be marked on the edge of a 3 by 5 card, and for ellipses of intermediate size the diallist will use his ingenuity.

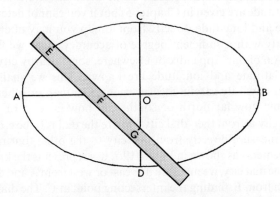

FIGURE 19.6 Laying out an ellipse with a trammell.

Setting the Dial in Place. Even though a sundial has been carefully and properly designed, it cannot tell the time correctly unless it is properly set in place. With portable dials this may be roughly accomplished with a compass to show the north and by hanging the instrument up to get it plumb. But a permanent sundial deserves reasonable care in giving it the proper location. It goes without saying that one should select a spot which is open to the sun during most of the day. If the instrument is a horizontal dial, the dial plate should be carefully leveled with a carpenter's level; or a vertical dial should be accurately plumbed with a plumb line or with a carpenter's level. The line showing 12 o'clock noon local apparent time should lie in the meridian. With vertical dials, this means that this hour line should be vertical. With horizontal dials it involves placing the hour line accurately in a north-south direction—"in the meridian." Several methods for determining the meridian were outlined in Chapter 3. If you are designing a dial to show mean time (one which includes longitude corrections such as those described in Chapters 2 and 5) the hour line of noon local apparent time will not appear; but in such cases it should be marked obscurely when laying out the design for use in orienting the finished dial. Longitude corrections can be built into any dial by precisely the same methods as those described in Chapter 5 for horizontal dials—by computing the position of the shadow at a few minutes before or after the hour by local apparent time, when the sun has the appropriate hour angle at the standard time meridian.

Finding Latitude and Longitude. Some brief directions for finding your latitude are given in Chapter 5; but if you cannot determine the latitude and longitude directly from a map you may often proceed indirectly with a sufficient degree of accuracy. First we find, from Table A.6 of our Appendix or elsewhere, some nearby city or town whose latitude and longitude are known. This we shall call the "known city." By careful measurement on a map or otherwise, we determine how far north or south, and how far east or west, the known city is from the "dial city" where the dial is to be erected. We do *not* measure directly from one city to the other, but make two measurements as shown in Figure 19.7. If *A* represents the known city and *B* the dial city, we measure due east or west from *A* and due north or south from *B*, finding the intersecting point at *C*. The dial city lies, in Figure 19.7, west of the known city by a distance *AC* and south of

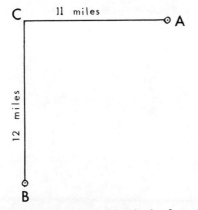

FIGURE 19.7 Finding the latitude and longitude of one city from those of another city close by.

the known city by a distance *CB*. We compare these distances with the scale of miles on the map, and find their lengths in miles. We then convert these mileages to degrees or minutes of latitude and longitude.

Suppose, for example, we want to know the latitude and longitude of Altoona, Pennsylvania. Table A.6 of our Appendix gives the location of nearby State College, Pennsylvania as 40°48′ N and 77°52′W. Measuring on an automobile road map, for want of a better source, we find that Altoona is 21 miles south and 29 miles west of State College. Each mile of north-south direction is the equivalent of about 0.868 minutes of latitude, so the 21 miles south are equivalent to about 18 minutes of latitude. Since Altoona is south of State College we subtract to find that the approximate latitude of Altoona is 40°48′ − 18′ = 40°30′ N. The adjustment for longitude is not quite so simple. The east-west mileage corresponding to one minute of longitude varies with the latitude, and our adustment factor is 0.868/cos latitude. Since we just estimated that the latitude of Altoona is 40°30′ our adjustment factor is 0.868/cos 40°30′ or 0.868/0.760 = 1.14, and each east-west mile is the equivalent of about 1.14 minutes of longitude at this latitude. Hence our 29 miles west are the equivalent of (29)(1.14) or about 33 minutes of longitude. Since Altoona is west of State College, we add to get 77°52′ + 33′ = 78°25′ as the approximate longitude of Altoona. Atlases give the actual coordinates of Altoona as 40°32′ N and 78°23′ W, and our estimate is plenty close enough for dialling purposes.

Appendix

HOW TO USE THE TABLES

Table A.1. The Equation of Time. This table shows the average amount by which local mean time differs from local apparent time at noon on each day of the year. If proper corrections have been made for longitude (see Chapter 2), the table shows the number of minutes and seconds by which the sundial is fast or slow compared with an accurate clock. For more accurate values, consult a current ephemeris; but the values tabulated here will not differ from the actual values by more than 10–15 seconds, and in most cases will be within 2–3 seconds of the actual values. For converting from mean to apparent time or vice versa our rules are:

(1) When the table states that the dial is "fast":
   (a) L.M.T. = L.A.T. − e
   (b) L.A.T. = L.M.T. + e
(2) When the table states that the dial is "slow":
   (a) L.M.T. = L.A.T. + e
   (b) L.A.T. = L.M.T. − e

Table A.2. Declinations of the Sun. This table shows how many degrees and minutes the sun is north or south of the celestial equator on any day of the year. When declinations are given with the negative sign (−) the sun is south of the equator. Otherwise the sun is north of the equator. The values given in the table are averages of the four-year leap-year cycle, and in any given year the actual value may differ slightly from that given here. Errors will be greatest at the times of the equinoxes and will practically disappear at the times of the solstices. They should never exceed about 10′–15′. For more precise values consult a current ephemeris or almanac.

Table A.3. The Zodiac. The ecliptic is a great circle along which the sun appears to move in its annual journey around the heavens. The movements of the major planets never carry them far from the ecliptic, and the zodiac is a band or belt 16° wide centered on the ecliptic within which the planets can be found. This belt is divided into twelve "signs," each covering 30° of celestial longitude. We commonly express the time of year by giving the month and day, but an alternative method is to state where the sun is on the ecliptic, or in which sign it appears. Thus we can say that it is November 18 or that the sun is at the 26° of Scorpio. Early sundials often carried lines by means of which one could tell in what sign the sun was located. These were, then, lines for showing the time of year, something like a modern calendar.

Table A.4. Conversion of Arc to Time. The sun appears to complete a circuit of the heavens in one day—360° in 24 hours. Thus each hour is the equivalent of 15° and each degree is the equivalent of 4 minutes. We often find it necessary or convenient to convert our data back and forth from degrees and minutes of arc to hours and minutes of time. Suppose, for example, that we are told that the longitude of Boise, Idaho is 116°12′ west. To convert this to its equivalent in time we note from Table A.4 that:

$$
\begin{aligned}
90° &= 6^h\ 00^m \\
26° &= 1^h\ 44^m \\
12′ &= \quad\ \ 00^m\ 48^s \\
\hline
\text{totals } 116°12′ &= 7^h\ 44^m\ 48^s
\end{aligned}
$$

Table A.5. Conversion of Time to Arc. This table facilitates the solution of such problems as: What is the sun's hour angle at $2^h\ 15^m\ 27^s$ P.M.? From Table A.5 we find:

$$
\begin{aligned}
2^h &= \quad\ \ 30° \\
15^m &= \quad\ \ 3°45′ \\
27^s &= \quad\ \ \ 6′45″ \\
\hline
\text{totals } 2^h\ 15^m\ 27^s &= 33°51′45″
\end{aligned}
$$

Table A.6. Latitudes, Longitudes and Standard Time Zones. Since the communities listed in this table are all in the United States, the latitudes are all north and the longitudes are all west. Time zones are numbered to show how many hours must be added to the local standard time to find the corresponding time at Greenwich, England. Over a quarter of the U.S. population lives in the communities here listed. The time zones are:

zone number	standard time	standard meridian
4	Atlantic	60° W
5	Eastern	75° W
6	Central	90° W
7	Mountain	105° W
8	Pacific	120° W
9	Yukon	135° W
10	Alaska-Hawaii	150° W
11	Bering	165° W

Table A.7. Earliest Sunrise and Latest Sunset. In laying out a sundial one includes hour lines only for those hours when the sun can shine on the dial. The times of sunrise and sunset vary through the year and at different latitudes. In any latitude, the earliest sunrise and latest sunset occur late in June in the Northern Hemisphere. This table gives the times when the sun's upper limb will appear to be on the horizon after allowance has been made for atmospheric refraction. Although the table is drawn up for the Northern Hemisphere, it will serve for the Southern Hemisphere with minor error.

Table A.8. Sunset Times at Various Latitudes and Seasons. The time of sunrise or sunset varies with the latitude and the season. This table covers most latitudes of the United States and much of Europe, and gives the local apparent time of sunset at various times of year. For example, if we want to know the time of sunset at St. Louis, Missouri, on March 2, we note that the latitude is 38°40′ N (Table A.6) and the sun's declination is −7°26′ (Table A.2). Table A.8 tells us at once that the sun sets between 5:30 and 5:40 P.M. local apparent time, and interpolation in the table would tell us that the time is more nearly 5:56 P.M. This would be corrected for longitude and for the equation of time in accordance with the rules outlined in Chapter 2 if we wanted the Central Standard Time of sunset in St. Louis. Sunrise times will be about as far before noon as these sunset times are after noon. (March 2 sunrise at St. Louis about 5:56 before noon or about 6:04 A.M.) For southern latitudes the table can be used by reversing the sign of the solar declination. Thus in the Southern Hemisphere we would use the table as though the sun's declination on March 2 were $+7°26'$ instead of $-7°26'$.

Table A.9. This table is used for drawing the hour lines on horizontal dials or on vertical direct south dials in a unit square as explained in Chapters 5 and 6 on pages 40 and 53. Measurements followed by the letter "v" are taken vertically on the sides of the unit square; others are taken horizontally on the top of the square for horizontal dials and on the bottom for vertical ones. Where necessary, interpolate for latitude to the nearest tenth of a degree, corresponding to about 7 miles on the earth's surface.

Table A.10. This table is used for laying out the hour lines on horizontal or vertical direct south dials in various latitudes. The table gives the angles which the hour lines make with the 12-o'clock or meridian line. Latitudes for horizontal dials appear across the top of the table—for vertical direct south dials across the bottom. Where necessary, interpolate for latitude to the nearest tenth of a degree.

Tables A.11 and A.12. These tables reduce the computation in laying out reflected ceiling dials. Their use is explained in Chapter 14. All figures in these tables are in terms of the vertical distance from the mirror to the ceiling, and must be multiplied by that distance to convert them to the actual measurements in use. Thus if the vertical distance from mirror to ceiling is 38 inches, and if our latitude is 35°, these tables tell us that the equinoctial line will be $(38)(0.7002) = 26.6$ inches north of the point on the ceiling vertically above the mirror; that the dial center will be $(38)(2.128) = 80.9$ inches south of the first point just found (and hence $80.9 - 26.6 = 54.3$ inches outdoors south of the house); and that the hour line for 10:30 A.M. will cross the equinoctial line $(38)(0.5057) = 19.2$ inches from the junction of the meridian and the equinoctial line.

TABLE A.1
THE EQUATION OF TIME

This table shows for each day of the year the number of minutes and seconds by which a sun-dial is fast or slow as compared with an accurate clock which shows local mean time. For places not on a standard time meridian, further correction is necessary as explained in the text. The table shows average values, and may be in error by as much as 10–15 seconds in December and January of some years.

day	Jan.	Feb.	Mar.	Apr.	May	June	July	Aug.	Sept.	Oct.	Nov.	Dec.
—	slow	slow	slow	slow	fast	fast	slow	slow	slow	fast	fast	fast
1	3^m12^s	13^m33^s	12^m34^s	4^m08^s	2^m51^s	2^m25^s	3^m33^s	6^m16^s	$s0^m12^s$	10^m05^s	16^m20^s	11^m11^s
2	3 40	13 41	12 23	3 40	2 59	2 16	3 45	6 13	f0 07	10 24	16 22	10 49
3	4 08	13 48	12 11	3 32	3 06	2 06	3 57	6 09	0 26	10 43	16 23	10 26
4	4 36	13 55	11 58	3 14	3 12	1 56	4 08	6 04	0 45	11 02	16 23	10 02
5	5 03	14 01	11 45	2 57	3 18	1 46	4 19	5 59	1 05	11 20	16 22	9 38
—	slow	slow	slow	slow	fast	fast	slow	slow	fast	fast	fast	fast
6	5 30	14 06	11 31	2 40	3 23	1 36	4 29	5 53	1 25	11 38	16 20	9 13
7	5 57	14 10	11 17	2 23	3 27	1 25	4 39	5 46	1 45	11 56	16 18	8 48
8	6 23	14 14	11 03	2 06	3 31	1 14	4 49	5 39	2 05	12 13	16 15	8 22
9	6 49	14 16	10 48	1 49	3 35	1 03	4 58	5 31	2 26	12 30	16 11	7 56
10	7 14	14 18	10 33	1 32	3 38	0 51	5 07	5 23	2 47	12 46	16 06	7 29
—	slow	slow	slow	slow	fast	fast	slow	slow	fast	fast	fast	fast
11	7 38	14 19	10 18	1 16	3 40	0 39	5 16	5 14	3 08	13 02	16 00	7 02
12	8 02	14 20	10 02	1 00	3 42	0 27	5 24	5 05	3 29	13 18	15 53	6 34
13	8 25	14 19	9 46	0 44	3 44	0 15	5 32	4 55	3 50	13 33	15 46	6 06
14	8 48	14 18	9 30	0 29	3 44	f0 03	5 39	4 44	4 11	13 47	15 37	5 38
15	9 10	14 16	9 13	s0 14	3 44	s0 10	5 46	4 33	4 32	14 01	15 28	5 09
—	slow	slow	slow	change	fast	slow	slow	slow	fast	fast	fast	fast
16	9 32	14 13	8 56	f0 01	3 44	0 23	5 52	4 21	4 53	14 14	15 18	4 40
17	9 52	14 10	8 39	0 15	3 43	0 36	5 58	4 09	5 14	14 27	15 07	4 11
18	10 12	14 06	8 22	0 29	3 41	0 49	6 03	3 57	5 35	14 39	14 56	3 42
19	10 32	14 01	8 04	0 43	3 39	1 02	6 08	3 44	5 56	14 51	14 43	3 13
20	10 50	13 55	7 46	0 56	3 37	1 15	6 12	3 30	6 18	15 02	14 30	2 43
—	slow	slow	slow	fast	fast	slow	slow	slow	fast	fast	fast	fast
21	11 08	13 49	7 28	1 00	3 34	1 28	6 15	3 16	6 40	15 12	14 16	2 13
22	11 25	13 42	7 10	1 21	3 30	1 41	6 18	3 01	7 01	15 22	14 01	1 43
23	11 41	13 35	6 52	1 33	3 24	1 54	6 20	2 46	7 22	15 31	13 45	1 13
24	11 57	13 27	6 34	1 45	3 21	2 07	6 22	2 30	7 43	15 40	13 28	0 43
25	12 12	13 18	6 16	1 56	3 16	2 20	6 24	2 14	8 04	15 47	13 11	f0 13
—	slow	slow	slow	fast	fast	slow	slow	slow	fast	fast	fast	change
26	12 26	13 09	5 58	2 06	3 10	2 33	6 25	1 58	8 25	15 54	12 53	s0 17
27	12 39	12 59	5 40	2 16	3 03	2 45	6 25	1 41	8 46	16 01	12 34	0 47
28	12 51	12 48	5 21	2 26	2 56	2 57	6 24	1 24	9 06	16 06	12 14	1 16
29	13 03	12 42	5 02	2 35	2 49	3 09	6 23	1 07	9 26	16 11	11 54	1 45
30	13 14	xx xx	4 44	2 43	2 41	3 21	6 21	0 49	9 46	16 15	11 33	2 14
31	13 24	xx xx	4 26	x xx	2 33	x xx	6 19	0 31	x xx	16 18	xx xx	2 43
—	slow	slow	slow	fast	fast	slow	slow	slow	fast	fast	fast	slow

TABLE A.2
DECLINATIONS OF THE SUN

	Jan.	Feb.	Mar.	Apr.	May	June	July	Aug.	Sept.	Oct.	Nov.	Dec.
1	−23°04′	−17°20′	−7°49′	4°18′	14°54′	21°58′	23°09′	18°10′	8°30′	−2°57′	−14°14′	−21°43′
2	−22 59	−17 03	−7 26	4 42	15 12	22 06	23 05	17 55	8 09	−3 20	−14 34	−21 52
3	−22 54	−16 46	−7 03	5 05	15 30	22 14	23 01	17 40	7 47	−3 44	−14 53	−22 01
4	−22 48	−16 28	−6 40	5 28	15 47	22 22	22 56	17 24	7 25	−4 07	−15 11	−22 10
5	−22 42	−16 10	−6 17	5 51	16 05	22 29	22 51	17 08	7 03	−4 30	−15 30	−22 18
6	−22 36	−15 52	−5 54	6 13	16 22	22 35	22 45	16 52	6 40	−4 53	−15 48	−22 25
7	−22 28	−15 34	−5 30	6 36	16 39	22 42	22 39	16 36	6 18	−5 16	−16 06	−22 32
8	−22 21	−15 15	−5 07	6 59	16 55	22 47	22 33	16 19	5 56	−5 39	−16 24	−22 39
9	−22 13	−14 56	−4 44	7 21	17 12	22 53	22 26	16 02	5 33	−6 02	−16 41	−22 46
10	−22 05	−14 37	−4 20	7 43	17 27	22 58	22 19	15 45	5 10	−6 25	−16 58	−22 52
11	−21 56	−14 18	−3 57	8 07	17 43	23 02	22 11	15 27	4 48	−6 48	−17 15	−22 57
12	−21 47	−13 58	−3 33	8 28	17 59	23 07	22 04	15 10	4 25	−7 10	−17 32	−23 02
13	−21 37	−13 38	−3 10	8 50	18 14	23 11	21 55	14 52	4 02	−7 32	−17 48	−23 07
14	−21 27	−13 18	−2 46	9 11	18 29	23 14	21 46	14 33	3 39	−7 55	−18 04	−23 11
15	−21 16	−12 58	−2 22	9 33	18 43	23 17	21 37	14 15	3 16	−8 18	−18 20	−23 14
16	−21 06	−12 37	−1 59	9 54	18 58	23 20	21 28	13 56	2 53	−8 40	−18 35	−23 17
17	−20 54	−12 16	−1 35	10 16	19 11	23 22	21 18	13 37	2 30	−9 02	−18 50	−23 20
18	−20 42	−11 55	−1 11	10 37	19 25	23 24	21 08	13 18	2 06	−9 24	−19 05	−23 22
19	−20 30	−11 34	−0 48	10 58	19 38	23 25	20 58	12 59	1 43	−9 45	−19 19	−23 24
20	−20 18	−11 13	−0 24	11 19	19 51	23 26	20 47	12 39	1 20	−10 07	−19 33	−23 25
21	−20 05	−10 52	0 00	11 39	20 04	23 26	20 36	12 19	0 57	−10 29	−19 47	−23 26
22	−19 52	−10 30	0 24	12 00	20 16	23 26	20 24	11 59	0 33	−10 50	−20 00	−23 26
23	−19 38	−10 08	0 47	12 20	20 28	23 26	20 12	11 39	0 10	−11 12	−20 13	−23 26
24	−19 24	−9 46	1 11	12 40	20 39	23 25	20 00	11 19	−0 14	−11 33	−20 26	−23 26
25	−19 10	−9 24	1 35	13 00	20 50	23 24	19 47	10 58	−0 37	−11 54	−20 38	−23 25
26	−18 55	−9 02	1 58	13 19	21 01	23 23	19 34	10 38	−1 00	−12 14	−20 50	−23 23
27	−18 40	−8 39	2 22	13 38	21 12	23 21	19 21	10 17	−1 24	−12 35	−21 01	−23 21
28	−18 25	−8 17	2 45	13 58	21 22	23 19	19 08	9 56	−1 47	−12 55	−21 12	−23 19
29	−18 09	−8 03	3 09	14 16	21 31	23 16	18 54	9 35	−2 10	−13 15	−21 23	−23 16
30	−17 53	xx xx	3 32	14 35	21 41	23 13	18 40	9 13	−2 34	−13 35	−21 33	−23 12

Solar declinations given in this table are the average for the 4 years of a leap-year cycle. Individual years may vary slightly from the figures given here, although at the solstices the errors are negligible, and at the equinoxes, when they are greatest, they do not exceed 8′ to 9′ of declination in leap years and in years next preceding leap years; and in the other years they never exceed about 3′ of declination even at the equinoxes.

TABLE A.3
THE ZODIAC

| name of the sign | | | symbol | when the sun enters the sign (approx.) | | | days in this sign | season of the sign |
Latin	English			date	declination	longitude		
Aries	Ram		♈	Mar. 20	0°00′	0°	30.5 ⎫	
Taurus	Bull		♉	Apr. 20	11°28.5′	30°	31.0 ⎬ 92.8	spring
Gemini	Twins		♊	May 21	20°09.3′	60°	31.3 ⎭	
Cancer	Crab		♋	June 21	23°26.6′	90°	31.5 ⎫	
Leo	Lion		♌	July 23	20°09.3′	120°	31.3 ⎬ 93.7	summer
Virgo	Virgin		♍	Aug. 23	11°28.5′	150°	30.9 ⎭	
Libra	Scales		♎	Sept. 23	0°00.0′	180°	30.4 ⎫	
Scorpio	Scorpion		♏	Oct. 23	−11°28.5′	210°	29.9 ⎬ 89.8	autumn
Sagittarius	Archer		♐	Nov. 22	−20°09.3′	240°	29.5 ⎭	
Capricornus	Goat		♑	Dec. 21	−23°26.6′	270°	29.4 ⎫	
Aquarius	Water bearer		♒	Jan. 20	−20°09.3′	300°	29.6 ⎬ 89.0	winter
Pisces	Fishes		♓	Feb. 18	−11°28.5′	330°	30.0 ⎭	

Dates for entering signs are averages and vary slightly from year to year. They are based on Eastern Standard Time. The reader will note that the sun is north of the equator for approximately 8 days longer than it is south of the equator, and that summer is the longest season in the northern hemisphere.

TABLE A.4
CONVERSION OF ARC TO TIME

arc: degrees minutes	time: h m m s	arc: degrees minutes	time: h m m s	arc: degrees minutes	time: h m m s		arc: sec.	time: sec.	arc: sec.	time: sec.
0	0 00	30	2 00	60	4 00		0	0.00	30	2.00
1	0 04	31	2 04	61	4 04		1	0.07	31	2.07
2	0 08	32	2 08	62	4 08		2	0.13	32	2.13
3	0 12	33	2 12	63	4 12		3	0.20	33	2.20
4	0 16	34	2 16	64	4 16		4	0.27	34	2.27
5	0 20	35	2 20	65	4 20		5	0.33	35	2.33
6	0 24	36	2 24	66	4 24		6	0.40	36	2.40
7	0 28	37	2 28	67	4 28		7	0.47	37	2.47
8	0 32	38	2 32	68	4 32		8	0.53	38	2.53
9	0 36	39	2 36	69	4 36		9	0.60	39	2.60
10	0 40	40	2 40	70	4 40		10	0.67	40	2.67
11	0 44	41	2 44	71	4 44		11	0.73	41	2.73
12	0 48	42	2 48	72	4 48		12	0.80	42	2.80
13	0 52	43	2 52	73	4 52		13	0.87	43	2.87
14	0 56	44	2 56	74	4 56		14	0.93	44	2.93
15	1 00	45	3 00	75	5 00		15	1.00	45	3.00
16	1 04	46	3 04	76	5 04		16	1.07	46	3.07
17	1 08	47	3 08	77	5 08		17	1.13	47	3.13
18	1 12	48	3 12	78	5 12		18	1.20	48	3.20
19	1 16	49	3 16	79	5 16		19	1.27	49	3.27
20	1 20	50	3 20	80	5 20		20	1.33	50	3.33
21	1 24	51	3 24	81	5 24		21	1.40	51	3.40
22	1 28	52	3 28	82	5 28		22	1.47	52	3.47
23	1 32	53	3 32	83	5 32		23	1.53	53	3.53
24	1 36	54	3 36	84	5 36		24	1.60	54	3.60
25	1 40	55	3 40	85	5 40		25	1.67	55	3.67
26	1 44	56	3 44	86	5 44		26	1.73	56	3.73
27	1 48	57	3 48	87	5 48		27	1.80	57	3.80
28	1 52	58	3 52	88	5 52		28	1.87	58	3.87
29	1 56	59	3 56	89	5 56		29	1.93	59	3.93
30	2 00	60	4 00	90	6 00		30	2.00	60	4.00

In any pair of columns above, find the number of *degrees* of arc in the left-hand column and the corresponding number of hours and minutes of time opposite it in the right-hand column; *or* find the number of *minutes* of arc in the left-hand column and the corresponding number of minutes and seconds of time opposite it in the right-hand column.

In either pair of columns above, find the number of seconds of arc in the left-hand column and the corresponding number of seconds of time opposite it in the right-hand column. Time is given in seconds and hundredths.

TABLE A.5
CONVERSION OF TIME TO ARC

time in hours	arc in degrees	time: min. sec.	arc: d m m s	time: min. sec.	arc: d m m s
0	0				
1	15	0	0 00	30	7 30
2	30	1	0 15	31	7 45
3	45	2	0 30	32	8 00
4	60	3	0 45	33	8 15
		4	1 00	34	8 30
5	75				
6	90	5	1 15	35	8 45
7	105	6	1 30	36	9 00
8	120	7	1 45	37	9 15
9	135	8	2 00	38	9 30
		9	2 15	39	9 45
10	150				
11	165	10	2 30	40	10 00
12	180	11	2 45	41	10 15
13	195	12	3 00	42	10 30
14	210	13	3 15	43	10 45
		14	3 30	44	11 00
15	225				
16	240	15	3 45	45	11 15
17	255	16	4 00	46	11 30
18	270	17	4 15	47	11 45
19	285	18	4 30	48	12 00
		19	4 45	49	12 15
20	300				
21	315	20	5 00	50	12 30
22	330	21	5 15	51	12 45
23	345	22	5 30	52	13 00
24	360	23	5 45	53	13 15
		24	6 00	54	13 30
		25	6 15	55	13 45
		26	6 30	56	14 00
		27	6 45	57	14 15
		28	7 00	58	14 30
		29	7 15	59	14 45

In the small table directly above, find the number of hours of time in the left column, and directly opposite it in the right column find the corresponding number of degrees of arc. In the larger table at the right, find the number of minutes of time in the column at the left, and opposite it in the right-hand column the corresponding number of degrees and minutes of arc; *or* find the number of seconds of time in the left-hand column and the corresponding number of minutes and seconds of arc in the right-hand column.

TABLE A.6
LATITUDES, LONGITUDES AND STANDARD TIME ZONES OF SELECTED U.S. CITIES

	latitude	longitude	time zone
Akron, Ohio	41°04′	81°31′	5
Albany, N.Y.	42 40	73 49	5
Albuquerque, N. Mex.	35 05	106 38	7
Allentown, Pa.	40 37	75 30	5
Amarillo, Tex.	35 14	101 50	6
Amherst, Mass.	42 23	72 31	5
Anaheim, Calif.	33 50	117 56	8
Annapolis, Md.	38 59	76 30	5
Atlanta, Ga.	33 45	84 24	5
Augusta, Me.	44 17	69 50	5
Austin, Tex.	30 18	97 47	6
Baltimore, Md.	39 18	76 38	5
Baton Rouge, La.	30 30	91 10	6
Beaumont, Tex.	30 04	94 06	6
Berkeley, Calif.	37 53	122 17	8
Bismark, N.D.	46 50	100 18	6
Boise, Idaho	43 38	116 12	7
Boston, Mass.	42 20	71 05	5
Bridgeport, Conn.	42 12	73 12	5
Buffalo, N.Y.	42 52	78 55	5
Burlington, Vt.	44 28	73 14	5
Cambridge, Mass.	42 22	71 06	5
Camden, N.J.	39 52	75 07	5
Canton, Ohio	40 48	81 23	5
Carson City, Nev.	39 10	119 46	8
Charleston, W. Va.	38 23	81 40	5
Charlotte, N.C.	35 03	80 50	5
Chattanooga, Tenn.	35 02	85 18	5
Cheyenne, Wyo.	41 08	104 50	7
Chicago, Ill.	41 50	87 45	6
Cincinnati, Ohio	39 10	84 30	5
Cleveland, Ohio	41 30	81 41	5
Columbia, S.C.	34 00	81 00	5
Columbus, Ga.	32 28	84 59	5
Columbus, Ohio	39 59	83 03	5
Concord, N.H.	43 13	71 34	5
Corpus Christi, Tex.	27 47	97 26	6
Dallas, Tex.	32 47	96 48	6
Dawson, Alaska	64 04	139 24	9
Dayton, Ohio	39 45	84 10	5
Dearborn, Mich.	42 18	83 14	5
Denver, Colo.	39 45	105 00	7
Des Moines, Iowa	41 35	93 35	6
Detroit, Mich.	42 23	83 05	5
Dover, Del.	39 10	75 32	5
Duluth, Minn.	46 45	92 10	5

	latitude	longitude	time zone
Elizabeth, N.J.	40°40′	74°13′	5
El Paso, Tex.	31 45	106 30	7
Erie, Pa.	42 07	80 05	5
Evansville, Ind.	38 00	87 33	6
Fairbanks, Alaska	64 50	147 50	10
Farmington, Conn.	41 43	72 49	5
Flint, Mich.	43 03	83 40	5
Ft. Wayne, Ind.	41 05	85 08	6
Ft. Worth, Tex.	32 45	97 20	6
Frankfort, Ky.	38 11	84 53	6
Gary, Ind.	41 34	87 20	6
Glendale, Calif.	34 09	118 20	8
Grand Rapids, Mich.	42 57	86 40	5
Great Falls, Mont.	47 30	111 16	7
Greensboro, N.C.	36 03	79 50	5
Hammond, Ind.	41 37	87 30	6
Harrisburg, Pa.	40 17	76 54	5
Hartford, Conn.	41 45	72 42	5
Helena, Mont.	46 35	112 00	7
Honolulu, Hawaii	21 19	157 50	10
Houston, Tex.	29 45	95 25	6
Indianapolis, Ind.	39 45	86 10	6
Jackson, Miss.	32 20	90 11	6
Jackson Twp., Kans.	38 26	97 44	6
Jacksonville, Fla.	30 20	81 40	5
Jefferson City, Mo.	38 33	92 10	6
Jersey City, N.J.	40 44	74 04	5
Juneau, Alaska	58 20	134 20	8
Kansas City, Kans.	39 05	94 37	6
Kansas City, Mo.	39 02	94 33	6
Ketchican, Alaska	55 25	131 40	8
Knoxville, Tenn.	36 00	83 57	5
Lansing, Mich.	42 44	85 34	5
Lincoln, Mass.	42 25	71 17	5
Lincoln, Neb.	40 49	96 41	6
Little Rock, Ark.	34 42	34 46	6
Long Beach, Calif.	33 47	118 15	8
Los Angeles, Calif.	34 00	118 15	8
Lousville, Ky.	38 13	85 48	6
Lubbock, Tex.	33 35	101 53	6
Madison, Wis.	43°04′	89°22′	6
Manchester, N.H.	42 59	71 28	5

TABLE A.6 continued.

	latitude	longitude	time zone
Manhattan, Kans.	39 11	96 35	6
McPherson, Kans.	38 22	97 41	6
Memphis, Tenn.	35 10	90 00	6
Miami, Fla.	25 45	80 15	5
Milwaukee, Wis.	43 03	87 56	6
Minneapolis, Minn.	45 00	93 15	6
Mobile, Ala.	30 40	88 05	6
Montgomery, Ala.	33 22	86 20	6
Montpelier, Vt.	44 16	72 34	5
Nashville, Tenn.	36 10	86 50	6
Newark, N.J.	40 44	74 11	5
New Haven, Conn.	41 18	72 55	5
New Orleans, La.	30 00	90 03	6
Newport News, Va.	36 59	76 26	5
New York, N.Y.	40 40	73 50	5
Niagara Falls, N.Y.	43 06	79 04	5
Nome, Alaska	64 30	165 30	11
Norfolk, Va.	36 54	76 18	5
Northampton, Mass.	42 18	72 38	5
Oakland, Calif.	37 50	122 15	8
Oklahoma City, Okla.	35 38	97 33	6
Old Faithful Geyser, Wyo.	44 27	110 49	7
Old Saybrook, Conn.	41 17	72 22	5
Olympia, Wash.	47 03	122 53	8
Omaha, Neb.	41 15	96 00	6
Pasadena, Calif.	34 10	118 09	8
Paterson, N.J.	40 55	74 10	5
Peoria, Ill.	40 43	89 38	6
Philadelphia, Pa.	40 00	75 10	5
Phoenix, Ariz.	33 30	112 03	7
Pierre, S.D.	44 23	100 20	6
Pittsburgh, Pa.	40 26	80 00	5
Ponce, P.R.	18 01	66 36	4
Portland, Me.	43 41	70 18	5
Portland, Ore.	45 32	122 40	8
Portsmouth, Va.	36 50	76 20	5
Providence, R.I.	41 50	71 25	5
Raleigh, N.C.	35 46	78 39	5
Richmond, Va.	37 34	77 27	5
Rochester, N.Y.	43 12	77 37	5
Rockford, Ill.	42 16	89 06	6
Sacramento, Calif.	38 32	121 30	8
St. Louis, Mo.	38 40	90 15	6
St. Paul, Minn.	45 00	93 10	6
St. Petersburg, Fla.	27 45	82 40	5

	latitude	longitude	time zone
Salem, Oregon	44 57	123 01	8
Salt Lake City, Utah	40 45	111 55	7
San Antonio, Tex.	29 25	98 30	6
San Diego, Calif.	32 45	117 10	8
San Francisco, Calif.	37 45	122 27	8
San Jose, Calif.	37 20	121 55	8
San Juan, P.R.	18 29	66 08	4
Santa Ana, Calif.	33 44	117 54	8
Santa Fe, N. Mex.	35 41	105 57	7
Savannah, Ga.	32 04	81 07	5
Scranton, Pa.	41 25	75 40	5
Seattle, Wash.	47 35	122 20	8
Sheboygan Falls, Wis.	43 44	87 47	6
Shreveport, La.	32 30	93 46	6
South Bend, Ind.	41 40	86 15	6
Spokane, Wash.	47 40	117 25	8
Springfield, Ill.	39 49	89 39	6
Springfield, Mass.	42 07	72 35	5
State College, Pa.	40 48	77 52	5
Stillwater, Okla.	36 07	97 03	6
Storrs, Conn.	41 49	72 15	5
Syracuse, N.Y.	43 03	76 10	5
Tacoma, Wash.	47 15	122 27	8
Tallahassee, Fla.	30 26	84 19	5
Tampa, Fla.	27 58	82 38	5
Toledo, Ohio	41 40	83 35	5
Topeka, Kans.	39 02	95 41	6
Torrance, Calif.	33 50	118 20	8
Trenton, N.J.	40 15	74 43	5
Tucson, Ariz.	32 15	110 57	7
Tulsa, Okla.	36 07	95 58	6
University Park, Pa.	40 48	77 52	5
Utica, N.Y.	43 06	75 15	5
Washington, D.C.	38 55	77 00	5
Waterbury, Conn.	41 33	73 03	5
Wichita, Kans.	37 43	97 20	6
Wichita Falls, Tex.	33 55	98 30	6
Winston-Salem· N.C.	36 05	80 18	5
Wolfeboro, N.H.	43 36	71 14	5
Worcester, Mass.	42 17	71 48	5
Yonkers, N.Y.	40 56	73 54	5
Youngstown, Ohio	41 05	80 40	5

TABLE A.7
APPROXIMATE LOCAL APPARENT TIME OF EARLIEST SUNRISE AND LATEST
SUNSET AT VARIOUS LATITUDES

latitude	sunrise (A.M.)	sunset (P.M.)	latitude	earliest sunrise (A.M.)	latest sunset (P.M.)
0°	5:53	6:10	35	4:45	7:18
1	5:52	6:12	36	4:43	7:21
2	5:50	6:13	37	4:40	7:24
3	5:49	6:15	38	4:38	7:27
4	5:47	6:17	39	4:34	7:30
5	5:46	6:18	40	4:30	7:33
6	5:44	6:20	41	4:27	7:37
7	5:43	6:21	42	4:23	7:40
8	5:41	6:23	43	4:20	7:43
9	5:40	6:25	44	4:17	7:47
10	5:38	6:26	45	4:13	7:51
11	5:36	6:28	46	4:09	7:55
12	5:35	6:30	47	4:05	8:00
13	5:33	6:32	48	4:00	8:04
14	5:31	6:33	49	3:55	8:08
15	5:29	6:35	50	3:50	8:13
16	5:28	6:37	51	3:44	8:19
17	5:26	6:39	52	3:39	8:24
18	5:24	6:41	53	3:33	8:30
19	5:22	6:43	54	3:27	8:36
20	5:20	6:44	55	3:20	8:43
21	5:18	6:46	56	3:13	8:51
22	5:16	6:48	57	3:05	8:58
23	5:14	6:50	58	2:56	9:07
24	5:12	6:52	59	2:46	9:18
25	5:10	6:54	60	2:35	9:28
26	5:07	6:56	61	2:21	9:39
27	5:05	6:58	62	2:07	9:53
28	5:02	7:00	63	1:50	10:10
29	5:00	7:03	64	1:29	10:31
30	4:58	7:05	65	0:58	11:02
31	4:55	7:07	66	Sun never sets in	
32	4:53	7:09	etc.	midsummer at lati-	
33	4:50	7:12		tude 66° and above.	
34	4:48	7:15			

TABLE A.5

LOCAL APPARENT TIME OF SUNSET IN VARIOUS LATITUDES AND AT VARIOUS SOLAR DECLINATIONS

north latitude

solar declination	24°	26°	28°	30°	32°	34°	36°	38°	40°	42°	44°	46°	48°	50°	52°	54°
−22°	5h19m	5h15m	5h10m	5h06m	5h02m	4h57m	4h52m	4h46m	4h41m	4h35m	4h28m	4h21m	4h13m	4h05m	3h55m	3h45m
−20	5 23	5 19	5 15	5 11	5 07	5 03	4 59	4 54	4 49	4 43	4 38	4 31	4 25	4 17	4 09	4 00
−18	5 27	5 24	5 20	5 17	5 13	5 09	5 05	5 01	4 57	4 52	4 47	4 41	4 35	4 29	4 22	4 14
−16	5 31	5 28	5 25	5 22	5 19	5 15	5 12	5 08	5 04	5 00	4 56	4 51	4 46	4 40	4 34	4 27
−14	5 35	5 32	5 30	5 27	5 24	5 21	5 18	5 15	5 12	5 08	5 04	5 00	4 56	4 51	4 46	4 40
−12	5 38	5 36	5 34	5 32	5 29	5 27	5 24	5 22	5 19	5 16	5 13	5 09	5 05	5 01	4 57	4 52
−10	5 42	5 40	5 38	5 37	5 35	5 33	5 31	5 28	5 26	5 23	5 21	5 18	5 15	5 11	5 08	5 04
−8	5 46	5 44	5 43	5 42	5 40	5 38	5 37	5 35	5 33	5 31	5 29	5 27	5 24	5 21	5 19	5 15
−6	5 49	5 48	5 47	5 46	5 45	5 44	5 42	5 41	5 40	5 38	5 37	5 35	5 33	5 31	5 29	5 27
−4	5 53	5 52	5 51	5 51	5 50	5 49	5 48	5 47	5 47	5 46	5 45	5 43	5 42	5 41	5 39	5 38
−2	5 56	5 56	5 56	5 55	5 55	5 55	5 54	5 54	5 53	5 53	5 52	5 52	5 51	5 50	5 50	5 49
0	6 00	6 00	6 00	6 00	6 00	6 00	6 00	6 00	6 00	6 00	6 00	6 00	6 00	6 00	6 00	6 00
2	6 04	6 04	6 04	6 05	6 05	6 05	6 06	6 06	6 07	6 07	6 08	6 08	6 09	6 10	6 10	6 11
4	6 07	6 08	6 09	6 09	6 10	6 11	6 12	6 13	6 13	6 14	6 15	6 17	6 18	6 19	6 21	6 22
6	6 11	6 12	6 13	6 14	6 15	6 16	6 18	6 19	6 20	6 22	6 23	6 25	6 27	6 29	6 31	6 33
8	6 14	6 16	6 17	6 19	6 20	6 22	6 23	6 25	6 27	6 29	6 31	6 33	6 36	6 39	6 41	6 45
10	6 18	6 20	6 22	6 23	6 25	6 27	6 29	6 32	6 34	6 37	6 39	6 42	6 45	6 49	6 52	6 56
12	6 22	6 24	6 26	6 28	6 31	6 33	6 36	6 38	6 41	6 44	6 47	6 51	6 55	6 59	7 03	7 08
14	6 25	6 28	6 30	6 33	6 36	6 39	6 42	6 45	6 48	6 52	6 56	7 00	7 04	7 09	7 14	7 20
16	6 29	6 32	6 35	6 38	6 41	6 45	6 48	6 52	6 56	7 00	7 04	7 09	7 14	7 20	7 26	7 33
18	6 33	6 36	6 40	6 43	6 47	6 51	6 55	6 59	7 03	7 08	7 13	7 19	7 25	7 31	7 38	7 46
20	6 37	6 41	6 45	6 49	6 53	6 57	7 01	7 06	7 11	7 17	7 22	7 29	7 35	7 43	7 51	8 00
22	6 41	6 45nn	6 50	6 54	6 58	7 03	7 08	7 14	7 19	7 25	7 32	7 39	7 47	7 55	8 05	8 15

This table is drawn up for use in northern latitudes, but it can be used in the corresponding southern latitudes by changing the sign of the solar declination. Thus the sun sets at 34° south latitude at the same time when the solar declination is −12° as it does in 34° north latitude when the solar declination is 12°. In either case sunset occurs at 6h33m P.M. The sun rises at the same number of hours and minutes before local apparent noon as it sets after local apparent noon. Thus in 42° N latitude when the sun has a declination of −18°, the sun sets at 4:52 P.M. and rises at 4h52m before noon, or at 7:08 A.M.

TABLE A.9

VALUES FOR DRAWING HOUR LINES IN A UNIT SQUARE ON HORIZONTAL OR VERTICAL DIRECT SOUTH DIALS IN LATITUDES FROM 15° TO 75°

latitude, horizontal	11:30 or 12:30	11:00 or 1:00	10:30 or 1:30	10:00 or 2:00	9:30 or 2:30	9:00 or 3:00	8:30 or 3:30	8:00 or 4:00	7:30 or 4:30	7:00 or 5:00	6:30 or 5:30	latitude, vertical south
15°	.0341	.0694	.1072	.1494	.1986	.2588	.3373	.4483	.8002v	.5176v	.2543v	75°
16	.0363	.0739	.1143	.1593	.2117	.2759	.3596	.4779	.7506v	.4856v	.2388v	74
17	.0385	.0783	.1211	.1688	.2243	.2924	.3810	.9874v	.7084v	.4582v	.2251v	73
18	.0407	.0828	.1280	.1784	.2371	.3090	.4027	.9341v	.6702v	.4336v	.2130v	72
19	.0429	.0872	.1349	.1880	.2498	.3256	.4243	.8867v	.6361v	.4115v	.2022v	71
20	.0450	.0916	.1417	.1975	.2624	.3420	.4457	.8440v	.6055v	.3917v	.1925v	70
21	.0472	.0960	.1484	.2069	.2750	.3584	.4670	.8055v	.5779v	.3738v	.1837v	69
22	.0493	.1004	.1552	.2163	.2874	.3746	.4882	.7706v	.5528v	.3576v	.1757v	68
23	.0514	.1047	.1618	.2256	.2998	.3907	.9819v	.7388v	.5300v	.3429v	.1685v	67
24	.0535	.1090	.1677	.2348	.3121	.4067	.9433v	.7097v	.5092v	.3294v	.1618v	66
25	.0556	.1132	.1751	.2440	.3243	.4226	.9078v	.6831v	.4900v	.3170v	.1558v	65
26	.0577	.1174	.1816	.2531	.3364	.4384	.8752v	.6585v	.4724v	.3056v	.1502v	64
27	.0598	.1216	.1881	.2621	.3492	.4540	.8451v	.6359v	.4562v	.2951v	.1450v	63
28	.0618	.1258	.1945	.2710	.3602	.4695	.8172v	.6149v	.4411v	.2854v	.1402v	62
29	.0638	.1299	.2008	.2799	.3720	.4848	.7914v	.5954v	.4272v	.2763v	.1358v	61
30	.0658	.1340	.2071	.2887	.3837	.5000	.7673v	.5774v	.4142v	.2680v	.1317v	60
31	.0678	.1380	.2133	.2974	.3952	.9708v	.7449v	.5605v	.4021v	.2601v	.1278v	59
32	.0698	.1420	.2195	.3060	.4066	.9435v	.7240v	.5448v	.3908v	.2528v		
33	.0717	.1459	.2256	.3144	.4179	.9180v	.7044v	.5300v	.3803v	.2459v	.1209v	57
34	.0736	.1498	.2316	.3228	.4291	.8942v	.6861v	.5162v	.3704v	.2396v	.1177v	56
35	.0755	.1537	.2376	.3312	.4401	.8717v	.6689v	.5033v	.3611v	.2336v	.1148v	55
36	.0774	.1575	.2435	.3394	.4510	.8506v	.6527v	.4911v	.3523v	.2279v	.1120v	54
37	.0792	.1613	.2493	.3475	.4618	.8308v	.6375v	.4797v	.3441v	.2226v	.1094v	53
38	.0811	.1650	.2550	.3554	.4724	.8121v	.6232v	.4689v	.3364v	.2176v	.1069v	52
39	.0828	.1686	.2607	.3633	.4829	.7945v	.6096v	.4587v	.3291v	.2129v	.1046v	51
40	.0846	.1722	.2662	.3711	.4932	.7779v	.5969v	.4491v	.3222v	.2084v	.1024v	50

43	.0897	.1827	.2825	.3937	.9554v	.7332v	.5626v	.4233v	.3037v	.1964v	.0965v	47
44	.0915	.1861	.2877	.4011	.9380v	.7198v	.5523v	.4156v	.2981v	.1929v	.0948v	46
45	.0931	.1895	.2929	.4083	.9215v	.7071v	.5426v	.4082v	.2929v	.1895v	.0931v	45
46	.0947	.1927	.2980	.4153	.9059v	.6951v	.5334v	.4013v	.2879v	.1862v	.0915v	44
47	.0963	.1960	.3029	.4222	.8910v	.6837v	.5246v	.3947v	.2832v	.1832v	.0900v	43
48	.0978	.1991	.3078	.4291	.8768v	.6728v	.5163v	.3885v	.2787v	.1803v	.0886v	42
49	.0994	.2022	.3126	.4357	.8634v	.6625v	.5084v	.3825v	.2744v	.1775v	.0872v	41
50	.1009	.2053	.3173	.4423	.8506v	.6527v	.5008v	.3768v	.2704v	.1749v	.0859v	40
51	.1023	.2082	.3219	.4487	.8385v	.6434v	.4937v	.3715v	.2665v	.1724v	.0847v	39
52	.1037	.2111	.3264	.4550	.8269v	.6345v	.4869v	.3663v	.2628v	.1700v	.0835v	38
53	.1051	.2140	.3308	.4611	.8159v	.6261v	.4804v	.3615v	.2593v	.1678v	.0824v	37
54	.1065	.2168	.3351	.4671	.8054v	.6180v	.4742v	.3568v	.2560v	.1657v	.0814v	36
55	.1078	.2195	.3393	.4729	.7955v	.6104v	.4684v	.3524v	.2528v	.1635v	.0804v	35
56	.1091	.2221	.3434	.4786	.7860v	.6031v	.4628v	.3482v	.2498v	.1616v	.0794v	34
57	.1104	.2247	.3474	.4842	.7770v	.5962v	.4575v	.3442v	.2469v	.1597v	.0785v	33
58	.1116	.2272	.3513	.4896	.7684v	.5896v	.4524v	.3404v	.2442v	.1580v	.0776v	32
59	.1129	.2297	.3550	.4949	.7602v	.5833v	.4476v	.3368v	.2416v	.1563v	.0768v	31
60	.1140	.2320	.3587	.5000	.7524v	.5774v	.4430v	.3333v	.2391v	.1547v	.0760v	30
61	.1151	.2344	.3623	.9902v	.7450v	.5717v	.4387v	.3301v	.2368v	.1532v	.0753v	29
62	.1162	.2366	.3657	.9808v	.7380v	.5663v	.4345v	.3270v	.2346v	.1517v	.0746v	28
63	.1173	.2387	.3691	.9720v	.7313v	.5612v	.4306v	.3240v	.2324v	.1504v	.0739v	27
64	.1183	.2408	.3723	.9635v	.7250v	.5563v	.4269v	.3212v	.2304v	.1491v	.0732v	26
65	.1193	.2428	.3754	.9555v	.7190v	.5517v	.4233v	.3185v	.2285v	.1478v	.0726v	25
66	.1203	.2448	.3784	.9480v	.7133v	.5473v	.4200v	.3160v	.2267v	.1466v	.0721v	24
67	.1212	.2467	.3813	.9408v	.7079v	.5432v	.4168v	.3136v	.2250v	.1455v	.0715v	23
68	.1221	.2484	.3841	.9340v	.7028v	.5393v	.4138v	.3113v	.2234v	.1445v	.0710v	22
69	.1229	.2502	.3867	.9276v	.6980v	.5356v	.4110v	.3092v	.2218v	.1435v	.0705v	21
70	.1237	.2518	.3892	.9216v	.6934v	.5321v	.4083v	.3072v	.2204v	.1426v	.0701v	20
71	.1245	.2533	.3916	.9159v	.6892v	.5288v	.4058v	.3053v	.2190v	.1417v	.0696v	19
72	.1252	.2548	.3939	.9106v	.6851v	.5257v	.4034v	.3035v	.2178v	.1409v	.0692v	18
73	.1259	.2562	.3961	.9056v	.6814v	.5228v	.4012v	.3019v	.2166v	.1401v	.0688v	17
74	.1265	.2576	.3982	.9009v	.6779v	.5202v	.3991v	.3003v	.2155v	.1394v	.0685v	16
75	.1272	.2588	.4001	.8966v	.6746v	.5176v	.3972v	.2989v	.2144v	.1387v	.0681v	15

TABLE A.10
ANGLES OF HOUR LINES ON HORIZONTAL OR VERTICAL DIRECT
SOUTH DIALS, LATITUDES 15° TO 70°

degrees of latitude, horizontal dials

	15°	16°	17°	18°	19°	20°	21°	22°	23°	24°	25°	26°	27°	28°	29°	30°
noon	0°00'	0°00'	0°00'	0°00'	0°00'	0°00'	0°00'	0°00'	0°00'	0°00'	0°00'	0°00'	0°00'	0°00'	0°00'	0°00'
0ʰ15ᵐ	0 58	1 02	1 06	1 09	1 14	1 17	1 21	1 24	1 28	1 31	1 35	1 39	1 42	1 46	1 49	1 53
0 30	1 57	2 05	2 12	2 20	2 27	2 35	2 42	2 49	2 57	3 04	3 11	3 18	3 25	3 32	3 39	3 46
0 45	2 57	3 09	3 20	3 31	3 42	3 54	4 05	4 16	4 27	4 38	4 48	4 59	5 11	5 22	5 31	5 41
1 00	3 58	4 15	4 29	4 44	4 59	5 14	5 29	5 44	5 59	6 13	6 27	6 42	6 56	7 09	7 24	7 38
1 15	5 01	5 23	5 40	5 59	6 18	6 37	6 56	7 15	7 33	7 51	8 10	8 28	8 46	9 03	9 21	9 38
1 30	6 07	6 33	6 54	7 18	7 41	8 04	8 27	8 49	9 12	9 34	9 56	10 17	10 39	11 00	11 21	11 42
1 45	7 16	7 47	8 12	8 40	9 07	9 35	10 01	10 28	10 54	11 21	11 46	12 12	12 37	13 02	13 27	13 51
2 00	8 30	9 06	9 35	10 07	10 39	11 10	11 41	12 12	12 43	13 13	13 43	14 12	14 41	15 10	15 38	16 06
2 15	9 49	10 30	11 03	11 40	12 16	12 52	13 28	14 03	14 38	15 12	15 46	16 19	16 52	17 25	17 57	18 29
2 30	11 14	12 01	12 38	13 20	14 02	14 43	15 23	16 02	16 42	17 20	17 58	18 35	19 12	19 49	20 24	21 00
2 45	12 47	13 40	14 23	15 10	15 56	16 42	17 27	18 11	18 55	19 38	20 20	21 02	21 42	22 24	23 02	23 41
3 00	14 31	15 30	16 18	17 11	18 02	18 53	19 43	20 32	21 21	22 08	22 55	23 40	24 25	25 09	25 52	26 34
3 15	16 27	17 33	18 26	19 25	20 48	21 45	22 13	23 08	24 01	24 53	25 44	26 33	27 22	28 10	28 56	29 42
3 30	18 38	19 52	20 51	21 56	22 59	24 02	25 02	26 01	26 59	27 55	28 51	29 44	30 37	31 27	32 17	33 05
3 45	21 10	22 32	23 38	24 49	25 58	27 07	28 12	29 17	30 19	31 20	32 19	33 16	34 12	35 06	35 58	36 49
4 00	24 09	25 40	26 51	28 10	29 25	30 39	31 50	32 59	34 06	35 10	36 12	37 13	38 11	39 07	40 02	40 54
4 15	27 42	29 21	30 40	31 29	33 26	34 45	36 00	37 13	38 24	39 31	40 36	41 38	42 38	43 35	44 31	45 24
4 30	32 00	33 48	35 13	36 44	38 10	39 33	40 52	42 08	43 20	44 29	45 35	46 37	47 37	48 35	49 30	50 22
4 45	37 19	39 15	40 44	42 19	43 48	45 13	46 33	47 49	49 01	50 09	51 14	52 15	53 13	54 08	55 00	55 50
5 00	44 00	45 59	47 30	49 05	50 33	51 56	53 13	54 25	55 34	56 37	57 38	58 34	59 27	60 17	61 04	61 49
5 15	52 27	54 21	55 46	57 14	58 34	59 49	60 58	62 02	63 01	63 56	64 48	65 36	66 20	67 02	67 42	68 18
5 30	63 03	64 37	65 46	66 56	67 59	68 57	69 50	70 38	71 23	72 04	72 42	73 17	73 50	74 20	74 49	75 15
5 45	75 47	76 42	77 22	78 02	78 37	79 09	79 38	80 05	80 29	80 51	81 11	81 30	81 47	82 03	82 18	82 32
6 00	90 00	90 00	90 00	90 00	90 00	90 00	90 00	90 00	90 00	90 00	90 00	90 00	90 00	90 00	90 00	90 00
	75°	74°	73°	72°	71°	70°	69°	68°	67°	66°	65°	64°	63°	62°	61°	60°

degrees of latitude, horizontal dials

time	44°	43°	42°	41°	40°	39°	38°	37°	36°	35°	34°	33°	32°	31°
noon	0°00'	0°00'	0°00'	0°00'	0°00'	0°00'	0°00'	0°00'	0°00'	0°00'	0°00'	0°00'	0°00'	0°00'
0ʰ15ᵐ	2 36	2 34	2 31	2 28	2 25	2 22	2 19	2 16	2 12	2 09	2 06	2 03	1 59	1 56
0 30	5 14	5 08	5 02	4 56	4 50	4 44	4 38	4 32	4 26	4 20	4 13	4 07	3 59	3 53
0 45	7 52	7 44	7 35	7 26	7 17	7 08	6 59	6 50	6 40	6 31	6 21	6 11	6 01	5 51
1 00	10 33	10 21	10 10	9 58	9 46	9 34	9 22	9 10	8 57	8 44	8 31	8 18	8 05	7 52
1 15	13 16	13 02	12 48	12 33	12 19	12 04	11 48	11 33	11 17	11 11	10 45	10 29	10 12	9 55
1 30	16 03	15 47	15 29	15 12	14 55	14 37	14 18	14 00	13 41	13 22	13 03	12 43	12 23	12 02
1 45	18 55	18 35	18 16	17 56	17 35	17 15	16 53	16 32	16 10	15 48	15 25	15 02	14 39	14 15
2 00	21 51	21 29	21 07	20 45	20 22	19 58	19 34	19 10	18 45	18 19	17 54	17 27	17 01	16 34
2 15	24 54	24 30	24 05	23 40	23 15	22 49	22 22	21 55	21 27	20 58	20 29	20 00	19 30	18 59
2 30	28 04	27 38	27 11	26 43	26 15	25 47	25 17	24 47	24 17	23 45	23 14	22 41	22 08	21 34
2 45	31 21	30 53	30 24	29 55	29 25	28 54	28 22	27 50	27 16	26 42	26 08	25 32	24 56	24 18
3 00	34 47	34 18	33 47	33 16	32 44	32 11	31 37	31 03	30 27	29 50	29 13	28 35	27 55	27 15
3 15	38 23	37 53	37 21	36 48	36 14	35 40	35 04	34 28	33 50	33 11	32 32	31 51	31 09	30 25
3 30	42 09	41 38	41 05	40 32	39 57	39 22	38 44	38 07	37 27	36 47	36 05	35 22	34 38	33 52
3 45	46 07	45 35	45 02	44 28	43 54	43 17	42 39	42 01	41 20	40 39	39 56	39 11	38 25	37 37
4 00	50 16	49 45	49 13	48 39	48 04	47 28	46 50	46 12	45 31	44 49	44 05	43 20	42 33	41 44
4 15	54 38	54 08	53 37	53 04	52 30	51 55	51 18	50 40	50 00	49 19	48 36	47 50	47 03	46 14
4 30	59 12	58 44	58 15	57 44	57 12	56 39	56 04	55 28	54 50	54 10	53 29	52 45	51 59	51 12
4 45	63 58	63 32	63 06	62 38	62 10	61 40	61 08	60 36	59 59	59 23	58 45	58 04	57 21	56 37
5 00	68 55	68 33	68 11	67 47	67 22	66 56	66 29	66 00	65 29	64 58	64 24	63 48	63 11	62 31
5 15	74 01	73 44	73 27	73 08	72 48	72 28	72 06	71 43	71 18	70 53	70 25	69 56	69 25	68 53
5 30	79 16	79 05	78 52	78 39	78 26	78 11	77 56	77 40	77 23	77 04	76 45	76 25	76 03	75 40
5 45	84 37	84 31	84 24	84 18	84 11	84 03	83 55	83 47	83 38	83 29	83 19	83 08	82 57	82 45
6 00	90 00	90 00	90 00	90 00	90 00	90 00	90 00	90 00	90 00	90 00	90 00	90 00	90 00	90 00
	46°	47°	48°	49°	50°	51°	52°	53°	54°	55°	56°	57°	58°	59°

degrees of latitude, vertical south dials.

TABLE A.10 continued

degrees of latitude, horizontal dials

time	45°	46°	47°	48°	49°	50°	51°	52°	53°	54°	55°	56°	57°	58°	59°	60°
noon	0°00'	0°00'	0°00'	0°00'	0°00'	0°00'	0°00'	0°00'	0°00'	0°00'	0°00'	0°00'	0°00'	0°00'	0°00'	0°00'
0ʰ15ᵐ	2 39	2 42	2 45	2 47	2 50	2 52	2 55	2 57	3 00	3 02	3 04	3 07	3 09	3 11	3 13	3 15
0 30	5 19	5 25	5 30	5 35	5 41	5 45	5 51	5 55	6 00	6 05	6 09	6 14	6 18	6 22	6 26	6 30
0 45	8 00	8 09	8 17	8 25	8 32	8 40	8 47	8 55	9 02	9 09	9 15	9 22	9 29	9 35	9 41	9 46
1 00	10 44	10 55	11 05	11 16	11 26	11 36	11 46	11 55	12 05	12 14	12 23	12 31	12 40	12 48	12 46	13 04
1 15	13 30	13 43	13 57	14 10	14 22	14 35	14 47	14 59	15 10	15 21	15 33	15 43	15 54	16 04	16 14	16 23
1 30	16 20	16 35	16 51	17 07	17 22	17 36	17 51	18 05	18 18	18 32	18 45	18 57	19 09	19 21	19 33	19 44
1 45	19 14	19 32	19 50	20 08	20 25	20 42	20 58	21 14	21 30	21 45	22 00	22 14	22 28	22 42	22 55	23 08
2 00	22 12	22 33	22 53	23 13	23 33	23 51	24 10	24 28	24 45	25 02	25 19	25 35	25 50	26 05	26 20	26 34
2 15	25 18	25 40	26 31	26 25	26 46	27 06	27 27	27 46	28 05	28 24	28 42	28 59	29 16	29 32	29 48	30 03
2 30	28 29	28 54	29 18	29 42	30 05	30 27	30 49	31 10	31 30	31 50	32 09	32 28	32 46	33 03	33 20	33 36
2 45	31 49	32 15	32 41	33 06	33 30	33 54	34 17	34 39	35 01	35 22	35 42	36 01	36 20	36 38	36 56	37 13
3 00	35 16	35 42	36 11	36 37	37 03	37 27	37 51	38 14	38 37	38 59	39 20	39 40	39 59	40 18	40 36	40 53
3 15	38 53	39 22	39 49	40 17	40 43	41 08	41 33	41 56	42 20	42 42	43 03	43 24	43 43	44 02	44 21	44 38
3 30	42 40	43 09	43 37	44 05	44 32	44 57	45 22	45 46	46 09	46 31	46 52	47 13	47 33	47 52	48 10	48 27
3 45	46 37	47 07	47 35	48 03	48 29	48 54	49 19	49 42	50 05	50 27	50 48	51 08	51 27	51 46	52 04	52 21
4 00	50 46	51 15	51 43	52 10	52 35	53 00	53 24	53 46	54 09	54 30	54 50	55 09	55 28	55 45	56 03	56 19
4 15	55 06	55 34	56 00	56 26	56 50	57 13	57 36	57 57	58 18	58 38	59 01	59 19	59 36	59 49	60 05	60 20
4 30	59 38	60 04	60 29	60 52	61 15	61 36	61 57	62 16	62 35	62 53	63 11	63 27	63 43	63 58	64 13	64 26
4 45	64 21	64 44	65 06	65 27	65 47	66 06	66 24	66 42	66 58	67 14	67 30	67 44	67 58	68 11	68 24	68 36
5 00	69 15	69 34	69 53	70 10	70 27	70 43	70 59	71 13	71 28	71 41	71 53	72 06	72 17	72 28	72 39	72 49
5 15	74 17	74 32	74 47	75 01	75 14	75 27	75 39	75 50	76 01	76 11	76 21	76 31	76 39	76 48	76 56	77 04
5 30	79 27	79 38	79 48	79 57	80 06	80 15	80 24	80 31	80 38	80 46	80 52	80 59	81 05	81 11	81 16	81 21
5 45	84 42	84 48	84 53	84 58	85 02	85 07	85 11	85 15	85 19	85 22	85 26	85 29	85 32	85 35	85 38	85 40
6 00	90 00	90 00	90 00	90 00	90 00	90 00	90 00	90 00	90 00	90 00	90 00	90 00	90 00	90 00	90 00	90 00
	45°	44°	43°	42°	41°	40°	39°	38°	37°	36°	35°	34°	33°	32°	31°	30°

degrees of latitude, vertical south dials

TABLE A.10 continued

degrees of latitude, horizontal dials

time	70°	69°	68°	67°	66°	65°	64°	63°	62°	61°
noon 0h15m	0°00'	0°00'	0°00'	0°00'	0°00'	0°00'	0°00'	0°00'	0°00'	0°00'
0 30	3 32	3 30	3 28	3 27	3 26	3 24	3 22	3 21	3 19	3 17
0 45	7 03	7 01	6 58	6 55	6 51	6 48	6 45	6 41	6 38	6 34
	10 35	10 31	10 27	10 23	10 18	10 13	10 08	10 03	9 58	9 52
1 00	14 08	14 03	13 57	13 51	13 45	13 39	13 33	13 26	13 19	13 12
1 15	17 42	17 35	17 28	17 21	17 14	17 06	16 58	16 50	16 41	16 32
1 30	21 16	21 09	21 01	20 52	20 43	20 35	20 25	20 15	20 05	19 55
1 45	24 52	24 44	24 35	24 25	24 15	24 05	23 54	23 43	23 32	23 20
2 00	28 29	28 20	28 10	27 59	27 48	27 37	27 26	27 13	27 01	26 47
2 15	32 08	31 58	31 47	31 36	31 24	31 12	30 59	30 46	30 32	30 18
2 30	35 48	35 37	35 26	35 14	35 02	34 49	34 36	34 22	34 07	33 52
2 45	39 30	39 19	39 07	38 55	38 42	38 29	38 15	38 00	37 45	37 29
3 00	43 13	43 02	42 50	42 38	42 25	42 11	41 57	41 42	41 27	41 10
3 15	46 59	46 48	46 36	46 23	46 10	45 57	45 42	45 27	45 12	44 55
3 30	50 46	50 35	50 23	50 11	49 58	49 45	49 31	49 16	49 00	48 44
3 45	54 39	54 28	54 17	54 05	53 53	53 40	53 26	53 12	52 57	52 41
4 00	58 26	58 16	58 06	57 54	57 43	57 30	57 18	57 04	56 49	56 34
4 15	62 19	62 10	62 00	61 49	61 38	61 27	61 15	61 02	60 49	60 35
4 30	66 13	66 05	65 56	65 46	65 37	65 27	65 16	65 04	64 52	64 39
4 45	70 08	70 01	69 54	69 45	69 37	69 28	69 19	69 09	68 58	68 47
5 00	74 05	73 59	73 53	73 46	73 39	73 32	73 24	73 16	73 07	72 58
5 15	78 03	77 58	77 53	77 48	77 43	77 37	77 31	77 25	77 18	77 11
5 30	82 02	81 58	81 55	81 52	81 48	81 44	81 40	81 36	81 31	81 26
5 45	86 01	85 59	85 57	85 56	85 54	85 52	85 50	85 48	85 45	85 43
6 00	90 00	90 00	90 00	90 00	90 00	90 00	90 00	90 00	90 00	90 00
	20°	21°	22°	23°	24°	25°	26°	27°	28°	29°

degrees of latitude, vertical south dials

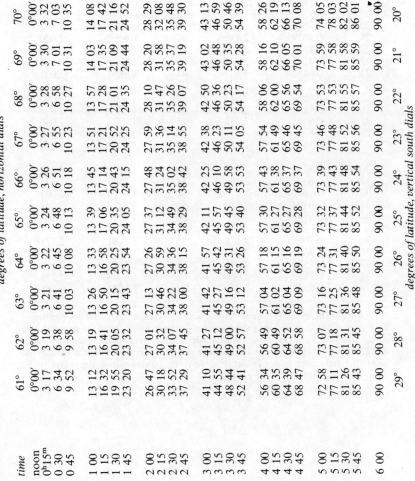

TABLE A.11
MULTIPLYING FACTORS FOR REFLECTED CEILING DIALS
IN LATITUDES 20° TO 59°[1]

latitude	multiplying factor for		latitude	multiplying factor for	
	equinox	center		equinox	center
20°	0.3640	3.112	40°	0.8391	2.031
21	0.3839	2.989	41	0.8693	2.020
22	0.4040	2.879	42	0.9004	2.011
23	0.4245	2.780	43	0.9325	2.005
24	0.4452	2.691	44	0.9657	2.001
25	0.4663	2.611	45	1.0000	2.000
26	0.4877	2.538	46	1.0355	2.001
27	0.5095	2.472	47	1.0724	2.005
28	0.5317	2.412	48	1.1106	2.011
29	0.5543	2.358	49	1.1504	2.020
30	0.5774	2.309	50	1.1918	2.031
31	0.6009	2.265	51	1.2349	2.045
32	0.6249	2.225	52	1.2799	2.061
33	0.6494	2.189	53	1.3270	2.081
34	0.6745	2.157	54	1.3764	2.103
35	0.7002	2.128	55	1.4281	2.128
36	0.7265	2.103	56	1.4826	2.157
37	0.7536	2.081	57	1.5399	2.189
38	0.7813	2.061	58	1.6003	2.225
39	0.8098	2.045	59	1.6643	2.265

[1] Values in the columns headed "equinox" are values of $\tan \phi$, and are multiplied by V to give the distance along the meridian from the point above the mirror to the equinoctial line. Values in the columns headed "center" are values of $\tan \phi + \cot \phi$, and are multiplied by V to give the distance along the meridian from the equinoctial line to the dial center. In both cases, V is the vertical distance from the mirror to the point vertically above it on the ceiling.

TABLE A.12

MULTIPLYING FACTORS FOR FINDING DISTANCES OF HOUR LINES FROM THE
EQUINOCTIAL ON REFLECTED CEILING DIALS LATITUDES 20° to 59°

hour of local apparent time

latitude	6:30 5:30	7:00 5:00	7:30 4:30	8:00 4:00	8:30 3:30	9:00 3:00	9:30 2:30	10:00 2:00	10:30 1:30	11:00 1:00	11:30 12:30
20°	8.0831	3.9715	2.5691	1.8432	1.3621	1.0642	0.8166	0.6144	0.4408	0.2851	0.1301
21	8.1362	3.9976	2.5860	1.8553	1.3710	1.0711	0.8219	0.6184	0.4437	0.2870	0.1410
22	8.1922	4.0251	2.6038	1.8681	1.3805	1.0785	0.8276	0.6227	0.4467	0.2890	0.1420
23	8.2516	4.0543	2.6227	1.8816	1.3905	1.0864	0.8336	0.6272	0.4500	0.2911	0.1430
24	8.3146	4.0853	2.6427	1.8960	1.4011	1.0946	0.8399	0.6320	0.4534	0.2933	0.1441
25	8.3809	4.1178	2.6638	1.9111	1.4123	1.1034	0.8466	0.6370	0.4570	0.2956	0.1453
26	8.4510	4.1523	2.6861	1.9271	1.4241	1.1126	0.8537	0.6424	0.4608	0.2981	0.1465
27	8.5249	4.1886	2.7096	1.9439	1.4365	1.1223	0.8612	0.6480	0.4649	0.3007	0.1478
28	8.6028	4.2269	2.7343	1.9617	1.4497	1.1326	0.8691	0.6539	0.4691	0.3035	0.1491
29	8.6846	4.2671	2.7603	1.9803	1.4634	1.1434	0.8773	0.6601	0.4736	0.3064	0.1505
30	8.7708	4.3094	2.7877	2.0000	1.4780	1.1547	0.8860	0.6667	0.4783	0.3094	0.1520
31	8.8614	4.3539	2.8165	2.0206	1.4932	1.1666	0.8952	0.6736	0.4832	0.3126	0.1536
32	8.9567	4.4008	2.8468	2.0424	1.5093	1.1792	0.9048	0.6808	0.4884	0.3160	0.1552
33	9.0569	4.4500	2.8787	2.0652	1.5262	1.1924	0.9149	0.6884	0.4939	0.3196	0.1570
34	9.1622	4.5017	2.9121	2.0893	1.5439	1.2062	0.9256	0.6964	0.4996	0.3232	0.1588
35	9.2728	4.5561	2.9473	2.1145	1.5625	1.2208	0.9367	0.7048	0.5057	0.3271	0.1607
36	9.3888	4.6131	2.9841	2.1409	1.5821	1.2361	0.9485	0.7136	0.5120	0.3312	0.1627
37	9.5109	4.6730	3.0229	2.1688	1.6027	1.2521	0.9608	0.7229	0.5186	0.3355	0.1648
38	9.6392	4.7361	3.0637	2.1980	1.6243	1.2690	0.9738	0.7327	0.5256	0.3400	0.1671
39	9.7739	4.8023	3.1066	2.2287	1.6470	1.2868	0.9874	0.7429	0.5330	0.3448	0.1694

TABLE A.12 continued

latitude	11:30 / 12:30	11:00 / 1:00	10:30 / 1:30	10:00 / 2:00	9:30 / 2:30	9:00 / 3:00	8:30 / 3:30	8:00 / 4:00	7:30 / 4:30	7:00 / 5:00	6:30 / 5:30
						hour of local apparent time					
40	0.1719	0.3498	0.5407	0.7537	1.0017	1.3054	1.6709	2.2611	3.1516	4.8719	9.9156
41	0.1744	0.3550	0.5488	0.7650	1.0167	1.3250	1.6959	2.2950	3.1989	4.9450	10.064
42	0.1772	0.3606	0.5575	0.7771	1.0328	1.3460	1.7228	2.3313	3.2495	5.0233	10.224
43	0.1800	0.3664	0.5664	0.7894	1.0492	1.3673	1.7501	2.3683	3.3010	5.1029	10.386
44	0.1830	0.3725	0.5758	0.8026	1.0667	1.3902	1.7794	2.4079	3.3562	5.1882	10.559
45	0.1862	0.3789	0.5858	0.8165	1.0852	1.4142	1.8101	2.4495	3.4142	5.2779	10.742
46	0.1900	0.3866	0.5977	0.8330	1.1072	1.4429	1.8472	2.4991	3.4835	5.3849	10.960
47	0.1930	0.3929	0.6074	0.8466	1.1251	1.4663	1.8768	2.5397	3.5400	5.4723	11.138
48	0.1968	0.4004	0.6190	0.8628	1.1468	1.4945	1.9129	2.5885	3.6080	5.5775	11.352
49	0.2007	0.4084	0.6314	0.8800	1.1696	1.5243	1.9510	2.6401	3.6799	5.6887	11.578
50	0.2048	0.4168	0.6444	0.8982	1.1937	1.5557	1.9866	2.6946	3.7559	5.8060	11.817
51	0.2092	0.4258	0.6582	0.9174	1.2193	1.5890	2.0339	2.7523	3.8363	5.9303	12.070
52	0.2138	0.4352	0.6728	0.9378	1.2464	1.6243	2.0790	2.8133	3.9214	6.0619	12.338
53	0.2188	0.4453	0.6883	0.9594	1.2750	1.6617	2.1268	2.8781	4.0116	6.2014	12.622
54	0.2240	0.4559	0.7047	0.9822	1.3055	1.7013	2.1771	2.9467	4.1073	6.3494	12.923
55	0.2296	0.4672	0.7222	1.0064	1.3378	1.7435	2.2315	3.0197	4.2091	6.5067	13.243
56	0.2354	0.4792	0.7407	1.0566	1.3722	1.7883	2.2889	3.0974	4.3174	6.6741	13.583
57	0.2417	0.4920	0.7605	1.0600	1.4089	1.8361	2.3501	3.1802	4.4327	6.8524	13.946
58	0.2484	0.5056	0.7816	1.0895	1.4480	1.8871	2.4154	3.2685	4.5559	7.0427	14.334
59	0.2556	0.5203	0.8042	1.1210	1.4899	1.9416	2.4851	3.3626	4.6875	7.2462	14.782

These are values of $(\tan \phi + \cot \phi)(\sin \phi)(\tan t)$

Bibliography

Bedos de Celles, Dom Francois. *La Gnomonique pratique*. Paris: 1790.

Bion, N. *De la Construction & des usages des cadrans solaires*. Paris: 1752.

Cross, Launcelot. *The Book of Old Sundials*. London: Foulis, 1914.

Diderot, Denis. *Encyclopedie ou dictionnaire raisonne des sciences*. Geneva: 1777.

Earle, Alice Morse. *Sun-dials and Roses of Yesterday*. London: Macmillan, 1922.

Gatty, Mrs. Alfred. *The Book of Sun-Dials*. 4th ed., enlarged and re-edited by H. K. F. Eden and Eleanor Lloyd. London: George Bell, 1900.

Ginzel, F. K. *Handbuch der mathematischen und technischen Chronologie*. Vol. III. Leipzig: 1914.

Green, Arthur Robert. *Sundials*. New York: Macmillan, 1926.

Holwell, John. *Clavis Horologiae*. London: 1712.

Hyatt, Alfred H. *A Book of Sundial Mottoes*. New York: Scott-Thaw, 1903.

Landon, Perceval. *Helio-tropes, or New Posies for Sundials*. London: Methuen, 1904.

Leybourn, William. *Dialling*. London: 1682.

Leadbetter, Charles. *Mechanick Dialling*. London: Caslon, 1773.

Mayall, Robert Newton, and Mayall, Margaret L. W. *Sundials*. Boston: Hale, Cushman & Flint, 1938.

Nautical Almanac Office of the U.S. Naval Observatory. *Ephemeris of the Sun, Polaris, and Other Selected Stars*. Washington, D.C.: U.S. Government Printing Office.

Price, Derek J. DeSolla. *Vistas in Astronomy*. Vol. 9. Oxford: Pergamon Press, 1968.

———. *Technology and Culture*, Vol. V, No. 1, 1964.

Prise, M. de la. *Cadrans solaires*. Caen: 1781.

Stebbins, Frederick A. "A Medieval Portable Sun-dial," *Journal of the Royal Astronomical Society of Canada*, April, 1961.

Sturmy, Capt. Samuel. *The Art of Dialling*. London: 1683.

United States Navy Department, Hydrographic Office. *Tables of Computed Altitude and Azimuth: U.S. Hydrographic Publication No. 214*. Washington, D.C.: U.S. Government Printing Office.

Index

A CATALOGUE OF SELECTED DOVER BOOKS
IN ALL FIELDS OF INTEREST

VISUAL ILLUSIONS: THEIR CAUSES, CHARACTERISTICS, AND APPLICATIONS, Matthew Luckiesh. Thorough description and discussion of optical illusion, geometric and perspective, particularly; size and shape distortions, illusions of color, of motion; natural illusions; use of illusion in art and magic, industry, etc. Most useful today with op art, also for classical art. Scores of effects illustrated. Introduction by William H. Ittleson. 100 illustrations. xxi + 252pp.

21530-X Paperbound $2.00

A HANDBOOK OF ANATOMY FOR ART STUDENTS, Arthur Thomson. Thorough, virtually exhaustive coverage of skeletal structure, musculature, etc. Full text, supplemented by anatomical diagrams and drawings and by photographs of undraped figures. Unique in its comparison of male and female forms, pointing out differences of contour, texture, form. 211 figures, 40 drawings, 86 photographs. xx + 459pp. 5⅜ x 8⅜.

21163-0 Paperbound $3.50

150 MASTERPIECES OF DRAWING, Selected by Anthony Toney. Full page reproductions of drawings from the early 16th to the end of the 18th century, all beautifully reproduced: Rembrandt, Michelangelo, Dürer, Fragonard, Urs, Graf, Wouwerman, many others. First-rate browsing book, model book for artists. xviii + 150pp. 8⅜ x 11¼.

21032-4 Paperbound $2.50

THE LATER WORK OF AUBREY BEARDSLEY, Aubrey Beardsley. Exotic, erotic, ironic masterpieces in full maturity: Comedy Ballet, Venus and Tannhauser, Pierrot, Lysistrata, Rape of the Lock, Savoy material, Ali Baba, Volpone, etc. This material revolutionized the art world, and is still powerful, fresh, brilliant. With *The Early Work,* all Beardsley's finest work. 174 plates, 2 in color. xiv + 176pp. 8⅛ x 11.

21817-1 Paperbound $3.75

DRAWINGS OF REMBRANDT, Rembrandt van Rijn. Complete reproduction of fabulously rare edition by Lippmann and Hofstede de Groot, completely reedited, updated, improved by Prof. Seymour Slive, Fogg Museum. Portraits, Biblical sketches, landscapes, Oriental types, nudes, episodes from classical mythology—All Rembrandt's fertile genius. Also selection of drawings by his pupils and followers. "Stunning volumes," *Saturday Review.* 550 illustrations. lxxviii + 552pp. 9⅛ x 12¼.

21485-0, 21486-9 Two volumes, Paperbound $10.00

THE DISASTERS OF WAR, Francisco Goya. One of the masterpieces of Western civilization—83 etchings that record Goya's shattering, bitter reaction to the Napoleonic war that swept through Spain after the insurrection of 1808 and to war in general. Reprint of the first edition, with three additional plates from Boston's Museum of Fine Arts. All plates facsimile size. Introduction by Philip Hofer, Fogg Museum. v + 97pp. 9⅜ x 8¼.

21872-4 Paperbound $2.50

GRAPHIC WORKS OF ODILON REDON. Largest collection of Redon's graphic works ever assembled: 172 lithographs, 28 etchings and engravings, 9 drawings. These include some of his most famous works. All the plates from *Odilon Redon: oeuvre graphique complet,* plus additional plates. New introduction and caption translations by Alfred Werner. 209 illustrations. xxvii + 209pp. 9⅛ x 12¼.

21966-8 Paperbound $4.50

DESIGN BY ACCIDENT; A BOOK OF "ACCIDENTAL EFFECTS" FOR ARTISTS AND DESIGNERS, James F. O'Brien. Create your own unique, striking, imaginative effects by "controlled accident" interaction of materials: paints and lacquers, oil and water based paints, splatter, crackling materials, shatter, similar items. Everything you do will be different; first book on this limitless art, so useful to both fine artist and commercial artist. Full instructions. 192 plates showing "accidents," 8 in color. viii + 215pp. 8⅜ x 11¼. 21942-9 Paperbound $3.75

THE BOOK OF SIGNS, Rudolf Koch. Famed German type designer draws 493 beautiful symbols: religious, mystical, alchemical, imperial, property marks, runes, etc. Remarkable fusion of traditional and modern. Good for suggestions of timelessness, smartness, modernity. Text. vi + 104pp. 6⅛ x 9¼. 20162-7 Paperbound $1.50

HISTORY OF INDIAN AND INDONESIAN ART, Ananda K. Coomaraswamy. An unabridged republication of one of the finest books by a great scholar in Eastern art. Rich in descriptive material, history, social backgrounds; Sunga reliefs, Rajput paintings, Gupta temples, Burmese frescoes, textiles, jewelry, sculpture, etc. 400 photos. viii + 423pp. 6⅜ x 9¾. 21436-2 Paperbound $5.00

PRIMITIVE ART, Franz Boas. America's foremost anthropologist surveys textiles, ceramics, woodcarving, basketry, metalwork, etc.; patterns, technology, creation of symbols, style origins. All areas of world, but very full on Northwest Coast Indians. More than 350 illustrations of baskets, boxes, totem poles, weapons, etc. 378 pp. 20025-6 Paperbound $3.00

THE GENTLEMAN AND CABINET MAKER'S DIRECTOR, Thomas Chippendale. Full reprint (third edition, 1762) of most influential furniture book of all time, by master cabinetmaker. 200 plates, illustrating chairs, sofas, mirrors, tables, cabinets, plus 24 photographs of surviving pieces. Biographical introduction by N. Bienenstock. vi + 249pp. 9⅞ x 12¾. 21601-2 Paperbound $5.00

AMERICAN ANTIQUE FURNITURE, Edgar G. Miller, Jr. The basic coverage of all American furniture before 1840. Individual chapters cover type of furniture—clocks, tables, sideboards, etc.—chronologically, with inexhaustible wealth of data. More than 2100 photographs, all identified, commented on. Essential to all early American collectors. Introduction by H. E. Keyes. vi + 1106pp. 7⅞ x 10¾. 21599-7, 21600-4 Two volumes, Paperbound $11.00

PENNSYLVANIA DUTCH AMERICAN FOLK ART, Henry J. Kauffman. 279 photos, 28 drawings of tulipware, Fraktur script, painted tinware, toys, flowered furniture, quilts, samplers, hex signs, house interiors, etc. Full descriptive text. Excellent for tourist, rewarding for designer, collector. Map. 146pp. 7⅞ x 10¾. 21205-X Paperbound $3.00

EARLY NEW ENGLAND GRAVESTONE RUBBINGS, Edmund V. Gillon, Jr. 43 photographs, 226 carefully reproduced rubbings show heavily symbolic, sometimes macabre early gravestones, up to early 19th century. Remarkable early American primitive art, occasionally strikingly beautiful; always powerful. Text. xxvi + 207pp. 8⅜ x 11¼. 21380-3 Paperbound $4.00

ALPHABETS AND ORNAMENTS, Ernst Lehner. Well-known pictorial source for decorative alphabets, script examples, cartouches, frames, decorative title pages, calligraphic initials, borders, similar material. 14th to 19th century, mostly European. Useful in almost any graphic arts designing, varied styles. 750 illustrations. 256pp. 7 x 10. 21905-4 Paperbound $4.00

PAINTING: A CREATIVE APPROACH, Norman Colquhoun. For the beginner simple guide provides an instructive approach to painting: major stumbling blocks for beginner; overcoming them, technical points; paints and pigments; oil painting; watercolor and other media and color. New section on "plastic" paints. Glossary. Formerly *Paint Your Own Pictures*. 221pp. 22000-1 Paperbound $1.75

THE ENJOYMENT AND USE OF COLOR, Walter Sargent. Explanation of the relations between colors themselves and between colors in nature and art, including hundreds of little-known facts about color values, intensities, effects of high and low illumination, complementary colors. Many practical hints for painters, references to great masters. 7 color plates, 29 illustrations. x + 274pp. 20944-X Paperbound $3.00

THE NOTEBOOKS OF LEONARDO DA VINCI, compiled and edited by Jean Paul Richter. 1566 extracts from original manuscripts reveal the full range of Leonardo's versatile genius: all his writings on painting, sculpture, architecture, anatomy, astronomy, geography, topography, physiology, mining, music, etc., in both Italian and English, with 186 plates of manuscript pages and more than 500 additional drawings. Includes studies for the Last Supper, the lost Sforza monument, and other works. Total of xlvii + 866pp. 7⅞ x 10¾. 22572-0, 22573-9 Two volumes, Paperbound $12.00

MONTGOMERY WARD CATALOGUE OF 1895. Tea gowns, yards of flannel and pillow-case lace, stereoscopes, books of gospel hymns, the New Improved Singer Sewing Machine, side saddles, milk skimmers, straight-edged razors, high-button shoes, spittoons, and on and on . . . listing some 25,000 items, practically all illustrated. Essential to the shoppers of the 1890's, it is our truest record of the spirit of the period. Unaltered reprint of Issue No. 57, Spring and Summer 1895. Introduction by Boris Emmet. Innumerable illustrations. xiii + 624pp. 8½ x 11⅝. 22377-9 Paperbound $8.50

THE CRYSTAL PALACE EXHIBITION ILLUSTRATED CATALOGUE (LONDON, 1851). One of the wonders of the modern world—the Crystal Palace Exhibition in which all the nations of the civilized world exhibited their achievements in the arts and sciences—presented in an equally important illustrated catalogue. More than 1700 items pictured with accompanying text—ceramics, textiles, cast-iron work, carpets, pianos, sleds, razors, wall-papers, billiard tables, beehives, silverware and hundreds of other artifacts—represent the focal point of Victorian culture in the Western World. Probably the largest collection of Victorian decorative art ever assembled— indispensable for antiquarians and designers. Unabridged republication of the Art-Journal Catalogue of the Great Exhibition of 1851, with all terminal essays. New introduction by John Gloag, F.S.A. xxxiv + 426pp. 9 x 12. 22503-8 Paperbound $5.00

A History of Costume, Carl Köhler. Definitive history, based on surviving pieces of clothing primarily, and paintings, statues, etc. secondarily. Highly readable text, supplemented by 594 illustrations of costumes of the ancient Mediterranean peoples, Greece and Rome, the Teutonic prehistoric period; costumes of the Middle Ages, Renaissance, Baroque, 18th and 19th centuries. Clear, measured patterns are provided for many clothing articles. Approach is practical throughout. Enlarged by Emma von Sichart. 464pp.   21030-8 Paperbound $3.50

Oriental Rugs, Antique and Modern, Walter A. Hawley. A complete and authoritative treatise on the Oriental rug—where they are made, by whom and how, designs and symbols, characteristics in detail of the six major groups, how to distinguish them and how to buy them. Detailed technical data is provided on periods, weaves, warps, wefts, textures, sides, ends and knots, although no technical background is required for an understanding. 11 color plates, 80 halftones, 4 maps. vi + 320pp. 6⅛ x 9⅛.   22366-3 Paperbound $5.00

Ten Books on Architecture, Vitruvius. By any standards the most important book on architecture ever written. Early Roman discussion of aesthetics of building, construction methods, orders, sites, and every other aspect of architecture has inspired, instructed architecture for about 2,000 years. Stands behind Palladio, Michelangelo, Bramante, Wren, countless others. Definitive Morris H. Morgan translation. 68 illustrations. xii + 331pp.   20645-9 Paperbound $3.00

The Four Books of Architecture, Andrea Palladio. Translated into every major Western European language in the two centuries following its publication in 1570, this has been one of the most influential books in the history of architecture. Complete reprint of the 1738 Isaac Ware edition. New introduction by Adolf Placzek, Columbia Univ. 216 plates. xxii + 110pp. of text. 9½ x 12¾.   21308-0 Clothbound $12.50

Sticks and Stones: A Study of American Architecture and Civilization, Lewis Mumford. One of the great classics of American cultural history. American architecture from the medieval-inspired earliest forms to the early 20th century; evolution of structure and style, and reciprocal influences on environment. 21 photographic illustrations. 238pp.   20202-X Paperbound $2.00

The American Builder's Companion, Asher Benjamin. The most widely used early 19th century architectural style and source book, for colonial up into Greek Revival periods. Extensive development of geometry of carpentering, construction of sashes, frames, doors, stairs; plans and elevations of domestic and other buildings. Hundreds of thousands of houses were built according to this book, now invaluable to historians, architects, restorers, etc. 1827 edition. 59 plates. 114pp. 7⅞ x 10¾   22236-5 Paperbound $4.00

Dutch Houses in the Hudson Valley Before 1776, Helen Wilkinson Reynolds. The standard survey of the Dutch colonial house and outbuildings, with constructional features, decoration, and local history associated with individual homesteads. Introduction by Franklin D. Roosevelt. Map. 150 illustrations. 469pp. 6⅝ x 9¼.   21469-9 Paperbound $5.00

THE ARCHITECTURE OF COUNTRY HOUSES, Andrew J. Downing. Together with Vaux's *Villas and Cottages* this is the basic book for Hudson River Gothic architecture of the middle Victorian period. Full, sound discussions of general aspects of housing, architecture, style, decoration, furnishing, together with scores of detailed house plans, illustrations of specific buildings, accompanied by full text. Perhaps the most influential single American architectural book. 1850 edition. Introduction by J. Stewart Johnson. 321 figures, 34 architectural designs. xvi + 560pp.
22003-6 Paperbound $5.00

LOST EXAMPLES OF COLONIAL ARCHITECTURE, John Mead Howells. Full-page photographs of buildings that have disappeared or been so altered as to be denatured, including many designed by major early American architects. 245 plates. xvii + 248pp. 7⅞ x 10¾. 21143-6 Paperbound $3.50

DOMESTIC ARCHITECTURE OF THE AMERICAN COLONIES AND OF THE EARLY REPUBLIC, Fiske Kimball. Foremost architect and restorer of Williamsburg and Monticello covers nearly 200 homes between 1620-1825. Architectural details, construction, style features, special fixtures, floor plans, etc. Generally considered finest work in its area. 219 illustrations of houses, doorways, windows, capital mantels. xx + 314pp. 7⅞ x 10¾. 21743-4 Paperbound $4.00

EARLY AMERICAN ROOMS: 1650-1858, edited by Russell Hawes Kettell. Tour of 12 rooms, each representative of a different era in American history and each furnished, decorated, designed and occupied in the style of the era. 72 plans and elevations, 8-page color section, etc., show fabrics, wall papers, arrangements, etc. Full descriptive text. xvii + 200pp. of text. 8⅜ x 11¼.
21633-0 Paperbound $5.00

THE FITZWILLIAM VIRGINAL BOOK, edited by J. Fuller Maitland and W. B. Squire. Full modern printing of famous early 17th-century ms. volume of 300 works by Morley, Byrd, Bull, Gibbons, etc. For piano or other modern keyboard instrument; easy to read format. xxxvi + 938pp. 8⅜ x 11.
21068-5, 21069-3 Two volumes, Paperbound $12.00

KEYBOARD MUSIC, Johann Sebastian Bach. Bach Gesellschaft edition. A rich selection of Bach's masterpieces for the harpsichord: the six English Suites, six French Suites, the six Partitas (Clavierübung part I), the Goldberg Variations (Clavierübung part IV), the fifteen Two-Part Inventions and the fifteen Three-Part Sinfonias. Clearly reproduced on large sheets with ample margins; eminently playable. vi + 312pp. 8⅛ x 11. 22360-4 Paperbound $5.00

THE MUSIC OF BACH: AN INTRODUCTION, Charles Sanford Terry. A fine, nontechnical introduction to Bach's music, both instrumental and vocal. Covers organ music, chamber music, passion music, other types. Analyzes themes, developments, innovations. x + 114pp. 21075-8 Paperbound $1.95

BEETHOVEN AND HIS NINE SYMPHONIES, Sir George Grove. Noted British musicologist provides best history, analysis, commentary on symphonies. Very thorough, rigorously accurate; necessary to both advanced student and amateur music lover. 436 musical passages. vii + 407 pp. 20334-4 Paperbound $4.00

JOHANN SEBASTIAN BACH, Philipp Spitta. One of the great classics of musicology, this definitive analysis of Bach's music (and life) has never been surpassed. Lucid, nontechnical analyses of hundreds of pieces (30 pages devoted to St. Matthew Passion, 26 to B Minor Mass). Also includes major analysis of 18th-century music. 450 musical examples. 40-page musical supplement. Total of xx + 1799pp.

(EUK) 22278-0, 22279-9 Two volumes, Clothbound $25.00

MOZART AND HIS PIANO CONCERTOS, Cuthbert Girdlestone. The only full-length study of an important area of Mozart's creativity. Provides detailed analyses of all 23 concertos, traces inspirational sources. 417 musical examples. Second edition. 509pp.

21271-8 Paperbound $4.50

THE PERFECT WAGNERITE: A COMMENTARY ON THE NIBLUNG'S RING, George Bernard Shaw. Brilliant and still relevant criticism in remarkable essays on Wagner's Ring cycle, Shaw's ideas on political and social ideology behind the plots, role of Leitmotifs, vocal requisites, etc. Prefaces. xxi + 136pp.

(USO) 21707-8 Paperbound $1.75

DON GIOVANNI, W. A. Mozart. Complete libretto, modern English translation; biographies of composer and librettist; accounts of early performances and critical reaction. Lavishly illustrated. All the material you need to understand and appreciate this great work. Dover Opera Guide and Libretto Series; translated and introduced by Ellen Bleiler. 92 illustrations. 209pp.

21134-7 Paperbound $2.00

BASIC ELECTRICITY, U. S. Bureau of Naval Personel. Originally a training course, best non-technical coverage of basic theory of electricity and its applications. Fundamental concepts, batteries, circuits, conductors and wiring techniques, AC and DC, inductance and capacitance, generators, motors, transformers, magnetic amplifiers, synchros, servomechanisms, etc. Also covers blue-prints, electrical diagrams, etc. Many questions, with answers. 349 illustrations. x + 448pp. 6½ x 9¼.

20973-3 Paperbound $3.50

REPRODUCTION OF SOUND, Edgar Villchur. Thorough coverage for laymen of high fidelity systems, reproducing systems in general, needles, amplifiers, preamps, loudspeakers, feedback, explaining physical background. "A rare talent for making technicalities vividly comprehensible," R. Darrell, *High Fidelity*. 69 figures. iv + 92pp.

21515-6 Paperbound $1.35

HEAR ME TALKIN' TO YA: THE STORY OF JAZZ AS TOLD BY THE MEN WHO MADE IT, Nat Shapiro and Nat Hentoff. Louis Armstrong, Fats Waller, Jo Jones, Clarence Williams, Billy Holiday, Duke Ellington, Jelly Roll Morton and dozens of other jazz greats tell how it was in Chicago's South Side, New Orleans, depression Harlem and the modern West Coast as jazz was born and grew. xvi + 429pp.

21726-4 Paperbound $3.95

FABLES OF AESOP, translated by Sir Roger L'Estrange. A reproduction of the very rare 1931 Paris edition; a selection of the most interesting fables, together with 50 imaginative drawings by Alexander Calder. v + 128pp. 6½x9¼.

21780-9 Paperbound $1.50

LAST AND FIRST MEN AND STAR MAKER, TWO SCIENCE FICTION NOVELS, Olaf Stapledon. Greatest future histories in science fiction. In the first, human intelligence is the "hero," through strange paths of evolution, interplanetary invasions, incredible technologies, near extinctions and reemergences. Star Maker describes the quest of a band of star rovers for intelligence itself, through time and space: weird inhuman civilizations, crustacean minds, symbiotic worlds, etc. Complete, unabridged. v + 438pp. (USO) 21962-3 Paperbound $3.00

THREE PROPHETIC NOVELS, H. G. WELLS. Stages of a consistently planned future for mankind. *When the Sleeper Wakes,* and *A Story of the Days to Come,* anticipate *Brave New World* and *1984,* in the 21st Century; *The Time Machine,* only complete version in print, shows farther future and the end of mankind. All show Wells's greatest gifts as storyteller and novelist. Edited by E. F. Bleiler. x + 335pp. (USO) 20605-X Paperbound $3.00

THE DEVIL'S DICTIONARY, Ambrose Bierce. America's own Oscar Wilde—Ambrose Bierce—offers his barbed iconoclastic wisdom in over 1,000 definitions hailed by H. L. Mencken as "some of the most gorgeous witticisms in the English language." 145pp. 20487-1 Paperbound $1.50

MAX AND MORITZ, Wilhelm Busch. Great children's classic, father of comic strip, of two bad boys, Max and Moritz. Also Ker and Plunk (Plisch und Plumm), Cat and Mouse, Deceitful Henry, Ice-Peter, The Boy and the Pipe, and five other pieces. Original German, with English translation. Edited by H. Arthur Klein; translations by various hands and H. Arthur Klein. vi + 216pp. 20181-3 Paperbound $2.00

PIGS IS PIGS AND OTHER FAVORITES, Ellis Parker Butler. The title story is one of the best humor short stories, as Mike Flannery obfuscates biology and English. Also included, That Pup of Murchison's, The Great American Pie Company, and Perkins of Portland. 14 illustrations. v + 109pp. 21532-6 Paperbound $1.50

THE PETERKIN PAPERS, Lucretia P. Hale. It takes genius to be as stupidly mad as the Peterkins, as they decide to become wise, celebrate the "Fourth," keep a cow, and otherwise strain the resources of the Lady from Philadelphia. Basic book of American humor. 153 illustrations. 219pp. 20794-3 Paperbound $2.00

PERRAULT'S FAIRY TALES, translated by A. E. Johnson and S. R. Littlewood, with 34 full-page illustrations by Gustave Doré. All the original Perrault stories—Cinderella, Sleeping Beauty, Bluebeard, Little Red Riding Hood, Puss in Boots, Tom Thumb, etc.—with their witty verse morals and the magnificent illustrations of Doré. One of the five or six great books of European fairy tales. viii + 117pp. 8⅛ x 11. 22311-6 Paperbound $2.00

OLD HUNGARIAN FAIRY TALES, Baroness Orczy. Favorites translated and adapted by author of the *Scarlet Pimpernel.* Eight fairy tales include "The Suitors of Princess Fire-Fly," "The Twin Hunchbacks," "Mr. Cuttlefish's Love Story," and "The Enchanted Cat." This little volume of magic and adventure will captivate children as it has for generations. 90 drawings by Montagu Barstow. 96pp. (USO) 22293-4 Paperbound $1.95

POEMS OF ANNE BRADSTREET, edited with an introduction by Robert Hutchinson. A new selection of poems by America's first poet and perhaps the first significant woman poet in the English language. 48 poems display her development in works of considerable variety—love poems, domestic poems, religious meditations, formal elegies, "quaternions," etc. Notes, bibliography. viii + 222pp.
22160-1 Paperbound $2.50

THREE GOTHIC NOVELS: THE CASTLE OF OTRANTO BY HORACE WALPOLE; VATHEK BY WILLIAM BECKFORD; THE VAMPYRE BY JOHN POLIDORI, WITH FRAGMENT OF A NOVEL BY LORD BYRON, edited by E. F. Bleiler. The first Gothic novel, by Walpole; the finest Oriental tale in English, by Beckford; powerful Romantic supernatural story in versions by Polidori and Byron. All extremely important in history of literature; all still exciting, packed with supernatural thrills, ghosts, haunted castles, magic, etc. xl + 291pp.
21232-7 Paperbound $3.00

THE BEST TALES OF HOFFMANN, E. T. A. Hoffmann. 10 of Hoffmann's most important stories, in modern re-editings of standard translations: Nutcracker and the King of Mice, Signor Formica, Automata, The Sandman, Rath Krespel, The Golden Flowerpot, Master Martin the Cooper, The Mines of Falun, The King's Betrothed, A New Year's Eve Adventure. 7 illustrations by Hoffmann. Edited by E. F. Bleiler. xxxix + 419pp. 21793-0 Paperbound $3.00

GHOST AND HORROR STORIES OF AMBROSE BIERCE, Ambrose Bierce. 23 strikingly modern stories of the horrors latent in the human mind: The Eyes of the Panther, The Damned Thing, An Occurrence at Owl Creek Bridge, An Inhabitant of Carcosa, etc., plus the dream-essay, Visions of the Night. Edited by E. F. Bleiler. xxii + 199pp. 20767-6 Paperbound $2.00

BEST GHOST STORIES OF J. S. LEFANU, J. Sheridan LeFanu. Finest stories by Victorian master often considered greatest supernatural writer of all. Carmilla, Green Tea, The Haunted Baronet, The Familiar, and 12 others. Most never before available in the U. S. A. Edited by E. F. Bleiler. 8 illustrations from Victorian publications. xvii + 467pp. 20415-4 Paperbound $3.00

MATHEMATICAL FOUNDATIONS OF INFORMATION THEORY, A. I. Khinchin. Comprehensive introduction to work of Shannon, McMillan, Feinstein and Khinchin, placing these investigations on a rigorous mathematical basis. Covers entropy concept in probability theory, uniqueness theorem, Shannon's inequality, ergodic sources, the E property, martingale concept, noise, Feinstein's fundamental lemma, Shanon's first and second theorems. Translated by R. A. Silverman and M. D. Friedman. iii + 120pp. 60434-9 Paperbound $2.00

SEVEN SCIENCE FICTION NOVELS, H. G. Wells. The standard collection of the great novels. Complete, unabridged. *First Men in the Moon, Island of Dr. Moreau, War of the Worlds, Food of the Gods, Invisible Man, Time Machine, In the Days of the Comet.* Not only science fiction fans, but every educated person owes it to himself to read these novels. 1015pp. (USO) 20264-X Clothbound $6.00

AGAINST THE GRAIN (A REBOURS), Joris K. Huysmans. Filled with weird images, evidences of a bizarre imagination, exotic experiments with hallucinatory drugs, rich tastes and smells and the diversions of its sybarite hero Duc Jean des Esseintes, this classic novel pushed 19th-century literary decadence to its limits. Full unabridged edition. Do not confuse this with abridged editions generally sold. Introduction by Havelock Ellis. xlix + 206pp. 22190-3 Paperbound $2.50

VARIORUM SHAKESPEARE: HAMLET. Edited by Horace H. Furness; a landmark of American scholarship. Exhaustive footnotes and appendices treat all doubtful words and phrases, as well as suggested critical emendations throughout the play's history. First volume contains editor's own text, collated with all Quartos and Folios. Second volume contains full first Quarto, translations of Shakespeare's sources (Belleforest, and Saxo Grammaticus), Der Bestrafte Brudermord, and many essays on critical and historical points of interest by major authorities of past and present. Includes details of staging and costuming over the years. By far the best edition available for serious students of Shakespeare. Total of xx + 905pp. 21004-9, 21005-7, 2 volumes, Paperbound $7.00

A LIFE OF WILLIAM SHAKESPEARE, Sir Sidney Lee. This is the standard life of Shakespeare, summarizing everything known about Shakespeare and his plays. Incredibly rich in material, broad in coverage, clear and judicious, it has served thousands as the best introduction to Shakespeare. 1931 edition. 9 plates. xxix + 792pp. 21967-4 Paperbound $4.50

MASTERS OF THE DRAMA, John Gassner. Most comprehensive history of the drama in print, covering every tradition from Greeks to modern Europe and America, including India, Far East, etc. Covers more than 800 dramatists, 2000 plays, with biographical material, plot summaries, theatre history, criticism, etc. "Best of its kind in English," *New Republic*. 77 illustrations. xxii + 890pp. 20100-7 Clothbound $10.00

THE EVOLUTION OF THE ENGLISH LANGUAGE, George McKnight. The growth of English, from the 14th century to the present. Unusual, non-technical account presents basic information in very interesting form: sound shifts, change in grammar and syntax, vocabulary growth, similar topics. Abundantly illustrated with quotations. Formerly *Modern English in the Making*. xii + 590pp. 21932-1 Paperbound $3.50

AN ETYMOLOGICAL DICTIONARY OF MODERN ENGLISH, Ernest Weekley. Fullest, richest work of its sort, by foremost British lexicographer. Detailed word histories, including many colloquial and archaic words; extensive quotations. Do not confuse this with the Concise Etymological Dictionary, which is much abridged. Total of xxvii + 830pp. 6½ x 9¼. 21873-2, 21874-0 Two volumes, Paperbound $7.90

FLATLAND: A ROMANCE OF MANY DIMENSIONS, E. A. Abbott. Classic of science-fiction explores ramifications of life in a two-dimensional world, and what happens when a three-dimensional being intrudes. Amusing reading, but also useful as introduction to thought about hyperspace. Introduction by Banesh Hoffmann. 16 illustrations. xx + 103pp. 20001-9 Paperbound $1.00

MATHEMATICAL PUZZLES FOR BEGINNERS AND ENTHUSIASTS, Geoffrey Mott-Smith. 189 puzzles from easy to difficult—involving arithmetic, logic, algebra, properties of digits, probability, etc.—for enjoyment and mental stimulus. Explanation of mathematical principles behind the puzzles. 135 illustrations. viii + 248pp.

20198-8 Paperbound $2.00

PAPER FOLDING FOR BEGINNERS, William D. Murray and Francis J. Rigney. Easiest book on the market, clearest instructions on making interesting, beautiful origami. Sail boats, cups, roosters, frogs that move legs, bonbon boxes, standing birds, etc. 40 projects; more than 275 diagrams and photographs. 94pp.

20713-7 Paperbound $1.00

TRICKS AND GAMES ON THE POOL TABLE, Fred Herrmann. 79 tricks and games— some solitaires, some for two or more players, some competitive games—to entertain you between formal games. Mystifying shots and throws, unusual caroms, tricks involving such props as cork, coins, a hat, etc. Formerly *Fun on the Pool Table*. 77 figures. 95pp.

21814-7 Paperbound $1.25

HAND SHADOWS TO BE THROWN UPON THE WALL: A SERIES OF NOVEL AND AMUSING FIGURES FORMED BY THE HAND, Henry Bursill. Delightful picturebook from great-grandfather's day shows how to make 18 different hand shadows: a bird that flies, duck that quacks, dog that wags his tail, camel, goose, deer, boy, turtle, etc. Only book of its sort. vi + 33pp. 6½ x 9¼.

21779-5 Paperbound $1.00

WHITTLING AND WOODCARVING, E. J. Tangerman. 18th printing of best book on market. "If you can cut a potato you can carve" toys and puzzles, chains, chessmen, caricatures, masks, frames, woodcut blocks, surface patterns, much more. Information on tools, woods, techniques. Also goes into serious wood sculpture from Middle Ages to present, East and West. 464 photos, figures. x + 293pp.

20965-2 Paperbound $2.50

HISTORY OF PHILOSOPHY, Julián Marias. Possibly the clearest, most easily followed, best planned, most useful one-volume history of philosophy on the market; neither skimpy nor overfull. Full details on system of every major philosopher and dozens of less important thinkers from pre-Socratics up to Existentialism and later. Strong on many European figures usually omitted. Has gone through dozens of editions in Europe. 1966 edition, translated by Stanley Appelbaum and Clarence Strowbridge. xviii + 505pp.

21739-6 Paperbound $3.50

YOGA: A SCIENTIFIC EVALUATION, Kovoor T. Behanan. Scientific but non-technical study of physiological results of yoga exercises; done under auspices of Yale U. Relations to Indian thought, to psychoanalysis, etc. 16 photos. xxiii + 270pp.

20505-3 Paperbound $2.50